The MAGICAL *World of* MATHEMATICS

Swagata Deb has done her PhD in Physics from IIT Delhi. She worked as a scientist for a couple of years till her daughter was born. After that she has been a mother, a teacher, a writer for children, a painter and a healer. Her published books include *Goopy Gyne Bagha Byne and Other Stories* which has been republished as a Puffin Classic. She is the co-author of *Why We Are Where We Are*, which was published by Scholastic in 2009.

The MAGICAL *World of* MATHEMATICS

Swagata Deb

RUPA

Published by
Rupa Publications India Pvt. Ltd. 2019
7/16, Ansari Road, Daryaganj
New Delhi 110002

Sales centres:

Allahabad Bengaluru Chennai
Hyderabad Jaipur Kathmandu
Kolkata Mumbai

ISBN: 978-93-5333-454-3

First impression 2019

10 9 8 7 6 5 4 3 2 1

CONTENTS

INTRODUCTION vii

1. THE MYSTERIES OF ZERO 1
2. HOW BIG IS INFINITY? 9
3. THE MAGIC NUMBER PI 16
4. WHAT ARE RANDOM NUMBERS? 22
5. FLIGHTS OF FANTASY: IMAGINARY NUMBERS 29
6. HOW MANY DIMENSIONS ARE THERE? 37
7. CREATING CHAOS IS SO EASY! 44
8. THE ENDLESS WORLD OF FRACTALS 50
9. THE PYRAMIDS ARE REALLY STRANGE! 56
10. WHY ACHILLES CAN'T EVER OUTRUN THE TORTOISE—THE PARADOXES OF MATHEMATICS 63
11. FIBONACCI NUMBERS: NATURE'S SERIES 69
12. 'THIS CAN'T BE HAPPENING!' THE MATHEMATICS OF IMPOSSIBLE OBJECTS 74
13. WHY YOU WON'T BE ABLE TO READ THIS MESSAGE 82
14. NUMBERS COULD BE ENTERTAINING! 90
15. IS ALL MUSIC ACTUALLY MATH? 95

16. ALL THAT THE PLANET NEEDS IS FOUR COLOURS 100

17. HOW THE WHOLE WORLD IS CONTAINED IN JUST TWO NUMBERS 106

18. WHY THE FINAL MYSTERY WILL ALWAYS BE BEYOND US: GÖDEL'S INCOMPLETENESS THEOREM 111

19. INDIA'S CONTRIBUTION TO MATHEMATICS 116

20. DO NUMBERS EXIST? 123

FURTHER READING 127

ACKNOWLEDGEMENTS 129

INTRODUCTION

I don't know if you enjoy Mathematics. I did, as a child. Today I teach Mathematics because I want to share my love for the subject with children. This book was written to introduce you to the beauty and scope of Mathematics.

Numbers are everywhere. From the shape of galaxies to the mysteries surrounding the pyramids, to the most beautiful music and art, there are patterns of numbers swirling around. Maths is also there in the abstract, in entertainment, linguistics, literature, mythology, nature, paradoxes, philosophy, psychology, science, astronomy, computers and even sociology.

Like many things in life, this book starts from 0. We discover how life would be almost impossible without this humble little number! Then we jump straight to the biggest number of them all—infinity. And find out that there are many infinities, and they get bigger and bigger! How can that be? Well, that's why Mathematics may seem so mysterious at times.

There are all sorts of numbers. There is that magic number pi. Computers have crashed, trying to calculate it perfectly, and there's even a day of the year named after it! Then there are random numbers—which are found in nature, and imaginary numbers—which are, well, imaginary, but are very useful if you want to build a big strong bridge!

All of us know about the pyramids. But are we aware of their strange Mathematics? The pyramids are situated at the exact centre of the landmass of the Earth. The curvature designed into the faces of the Great Pyramid exactly matches the radius of the Earth. How did the ancient Egyptians work all this out?

Mathematics abounds with paradoxes and puzzles. Take this statement: 'I am a liar.' If the statement is true, then I am not a liar. If that statement is false, then what does that make me? Yes, this is mind-boggling, isn't it? Mathematicians can also actually prove that the famous story of the hare and the tortoise can have a different ending: the hare can actually never outrun the tortoise in the race! Mathematicians also study 'impossible objects'. We live in a three-dimensional space. But some mathematicians work in spaces that have 12 or 16 or 24 dimensions!

And did you know that there are 'happy' numbers and 'unhappy' numbers and even 'perfect' numbers? And that a lot of great music and beautiful art could have a basis in Mathematics?

Computers and cell phones and the internet are extremely important parts of our lives. But they all know just two numbers—0 and 1. That's all. They haven't heard of any other number! So you can say that the whole world actually needs just two numbers!

Even among those of us who love Mathematics, how many know about India's contribution to this field? Indian mathematicians invented the zero, the decimal system, negative numbers, algebra and trigonometry, among many other things. They calculated the approximate value of pi, and worked out what we know today as the Pythagorean Theorem much before anyone else in the world. We must know about these men and remember them with the respect they deserve.

I end this book with two philosophical questions. One comes from what is called Godel's Incompleteness Theorem, which asks: Can we ever know everything? And the answer is: No. The second question is: Do numbers exist? Yes, there are lots of theories about that too, and lots of disagreements!

Whether or not we can ever know everything, and whether numbers really exist, what we do know is that knowledge can be both precious and fun. And that numbers can be mysterious, exciting and very cool.

I hope you enjoy reading this book as much as I enjoyed writing it!

1

THE MYSTERIES OF ZERO

Amongst all numbers, 0 holds a unique position. Let's try to understand why. 0 is neither a negative number nor a positive one. If you add 0 to a number, what do you get? You get the number itself! Subtracting 0 from a number also leaves the number intact. What happens when you multiply a number by 0? *Any* number multiplied by 0 is 0. Can you divide a number by 0? You can't. A number divided by 0 is undefined.

Early human beings, who had just learnt how to count, did not know about 0 at all, as it could not be observed in nature. Thus, 0 is not a natural number. 1, 2, 3 and the rest of the positive whole numbers are natural numbers. These numbers are easy to understand as they can be used to count objects. Negative numbers are also easy to grasp. If you say 'I have one apple' it means that you possess an apple. When you say 'I have minus one apple', it can mean that you owe someone an apple. 'Zero apples' means you have no apples. It is the same as saying 'zero unicorns'!

But the ease with which we write numbers today, particularly large ones, is all because of 0. Without 0, our lives would be very difficult!

Ancient civilizations had to really struggle in order to write down large numbers. The Babylonian method of recording quantities, about 5,000 years old, is one of the oldest number systems that we know of. The Babylonians made wedge-shaped marks on a clay tablet with the help of a blunt reed (a kind of grass). While we now use a base of 10, their number system had a base of 60. However, numbers less than 60 had a base of 10. So, unlike the way we count today, by 10s going up to 100 (that is, 10 × 10) and then 100s to 1000 (that is, 100 × 10), their 10s would go up to 60, and then the 60s would go up to 3600 (60 × 60), and so on.

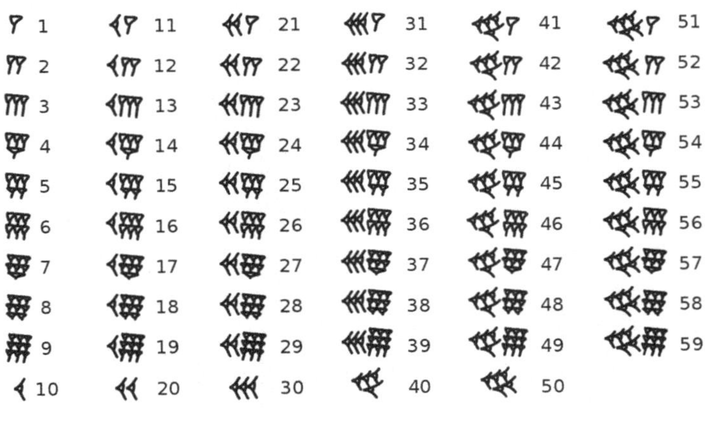

Figure 1.1: The oldest number system we know of

They developed a way of writing large numbers with the help of a few symbols. But they had no symbol for 0. So their numbers were difficult to interpret, and the same set of symbols would represent multiplication and division. For example, 21, 21 × 60, 21 × 60 × 60, and 21/60 would be represented using the same symbols. Depending on what the number was referring to—goats or ships or whatever, they would instinctively understand what number the symbols were referring to. Was it the number of goats someone owned or the price of a ship? This method must have made things very complicated!

The Egyptian method for recording quantities was also based on 10. There was a symbol for 1, 10 and each successive power of 10 (100, 1000 and so on). A distinct sign (hieroglyph) was used for each power of 10. But again, there was no symbol for 0. So the numbers were still difficult to read.

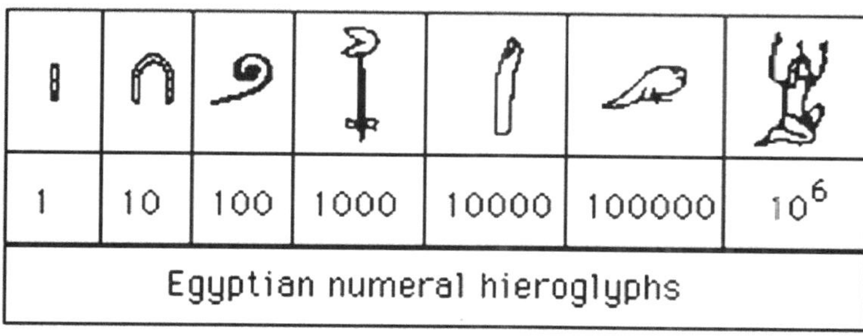

Figure 1.2: The numbers were difficult to read

The Mayans had an interesting numbering system that dates back to the fourth century CE. They used dots to represent numbers under 5. So 3 was three dots; 5 was a horizontal line; 6 was a horizontal line with a dot above it; 13 was two lines and three dots. This dot-dash system worked upto 19. Then, they used a shell to represent 0. 20 was one dot above a shell (I know, this does not sound very logical today, but that is only because we are used to writing numbers in a certain way). It was quite difficult to calculate numbers in the Mayan system. The Mayans certainly knew about the number 0, but they couldn't use it very effectively.

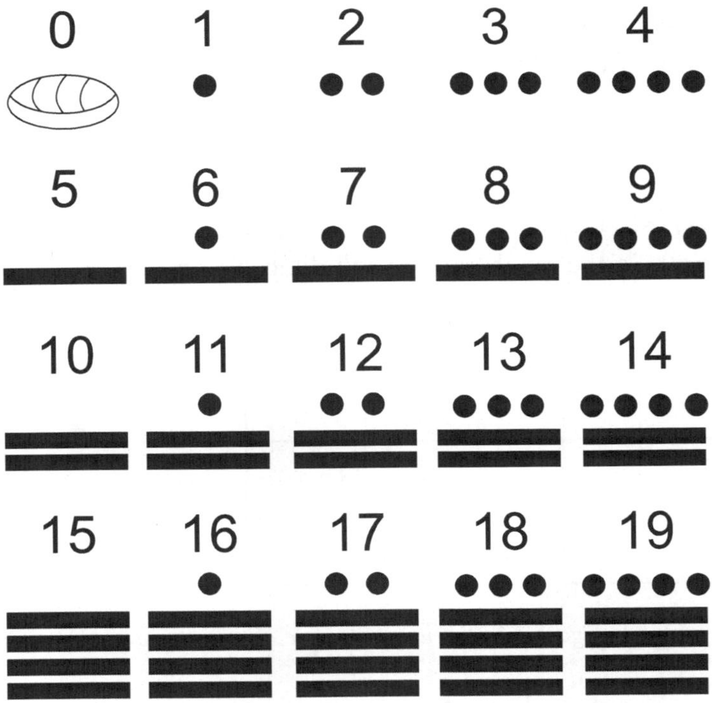

Figure 1.3: The Mayans had an interesting numbering system that dates back to the fourth century CE.

The zero, as a number and an idea, as we know it today, was discovered in India in the seventh century CE. Brahmagupta, a mathematician and astronomer, gave us the rules for using zero in arithmetic. He introduced many mathematical concepts, which seem very simple to us today. He also introduced the idea of negative numbers. His contribution of zero was important, as he formulated it as a number in its own right. Before him, many civilizations, the zero was just used as a 'placeholder'.

Brahmagupta used dots below numbers to indicate zero. These dots were referred to as 'shunya', which means empty, or 'kha', which means place. He was the first person to figure out that $1 + 0 = 1$, $1 - 0 = 1$ and $1 \times 0 = 0$. The only place where he went wrong was division by zero; he thought $1/0$ was 0.

Brahmagupta's work was adopted by Persian and Arab mathematicians, and this number system (which is what we use now in our daily life) came to be known as the Hindu-Arabic numeral system.

Six hundred years after Brahmagupta's wonderful discovery, the Hindu-Arabic numeral system—and, of course, zero—was introduced to the west by the great Italian mathematician Leonardo Fibonacci. He grew up with a North African education under the Moors, and he later travelled a lot around the Mediterranean coast. He must have met many merchants and learned of their systems of doing arithmetic. He soon realized the advantages of the Hindu-Arabic system over the others.

Before this, the Europeans were using Roman numerals, which were ill-suited for performing arithmetic operations.

For example:

CLXXIV added to XXVIII is CCII.

And XXVIII subtracted from CLXXIV is CXXXXVI.

Looking at the numbers, who could have possibly guessed what the result would be, or that the smallest-looking number was actually the biggest, and vice versa?

Introducing zero simplified arithmetic massively. The decimal system that we use today owes its existence to the fact that we have ten fingers and that 0 is recognized as a number. The base 10 is neither too large nor too small. If not for the discovery of zero and the way it simplified things, we would have had trouble memorizing symbols used for different digits.

Computers use the binary system. They count using just 0 and 1. So 1 is 01, 2 is 10, 3 is 11, 4 is 100. What we know as 32 is 1 followed by five zeroes! That seems very tedious, but it is actually the simplicity of the number system—using just two numbers. Thus, 0 allows computers to do very complicated things very fast.

Maybe if we had eight or twelve fingers, our numbering system would have been more useful. Can you think of a reason why?

From the decimal system, decimal fractions were derived (such as 0.8 or 127.95 or 6.1234982). These are very simple to write down and one can perform mathematical operations on them with great ease. Every real number (meaning fraction or integer) can be represented as a decimal fraction. So ¼ is 0.25 and 26 is 26.0. In a world where accuracy is of great importance, the decimal fraction is extremely useful. Calculus, which is a very advanced form of Mathematics, is based on reducing things to zero.

Calculus is a very powerful tool and essential for doing many practical things—for example, calculating volumes and surface areas of bodies with curved surfaces. Using calculus, one can divide the surface into extremely tiny bits and then add up the areas of all the tiny bits to obtain the total surface area. These tiny bits have to be as close to 0 as possible.

When we divide anything by 0, we get absurd results. Take the equation:

$$a = b \tag{1}$$

Multiplying by 'a' on both sides gives us:

$$a^2 = ab$$

When we add $(a^2 - 2ab)$ on both sides, we obtain:

$$2a^2 - 2ab = a^2 - ab$$

Both the terms in the LHS have 2 in common. We can write:

$$2(a^2 - ab) = a^2 - ab$$

Dividing both sides by $(a^2 - ab)$, we obtain

$$2 = 1$$

How odd is that! How did it happen? It happened because we divided both sides by $(a^2 - ab)$.

$$a^2 - ab = a(a - b)$$

$$\text{Since } a = b,\ (a - b) = 0.$$

Dividing by 0 gave us this absurd answer.

Zero is a valid and very necessary number, but it is unlike any other in many ways. In fact, older cultures had trouble accepting the concept of zero because it reminded them of the great void and chaos, which, according to Christian belief, separates the Earth from Hell!

In reality, zero has made all our lives much easier and better.

2

HOW BIG IS INFINITY?

You must have heard of the word 'infinity'. It's such a large number that we can't calculate or even conceive of it. You may be familiar with its symbol—an eight lying on its side instead of standing up (∞).

We know that infinity is a very big number. But how big is it? What happens if you add a number to infinity or subtract a number from infinity? What do you get when you multiply infinity by zero? And what is infinity minus infinity?

You must understand that infinity is not like other numbers. It is a concept used in Maths and Physics to make lives simpler for us. It originates from the Latin word 'infinitas', which denotes 'endlessness'.

Early Greeks were aware of the concept of infinity. But their idea of infinity was limited to a possibility of infinity rather than the actual existence of it. For example, they knew that there were more prime numbers than any given set of primes. But they did not take it one step further and say that the total number of prime numbers was infinity. Early Indians had a better idea of the infinite. They knew that if you remove a part from infinity or add a part to it, the answer would still remain infinite.

Let's see how we can obtain infinity from arithmetic. If we divide 1 by 0.1, we obtain 1/.01 = 100 and so on. We see that as we reduce the denominator, the answer becomes larger and larger. We say that when the denominator approaches zero, the answer approaches infinity.

Or imagine a line segment. How many points does it have? Strictly speaking, a point should not have any length, breadth or height. The length of a point would be too close to zero to actually measure. So the number of points in a line segment would be a finite length divided by a very short length—that is, almost zero. Therefore, we can say that there are infinite points in a line segment.

Calculus, which is a very advanced form of Mathematics, uses the concept of infinity. If we want to find the exact area of a sphere using calculus, we divide the surface of the sphere into extremely tiny bits, each with an area nearly equal to zero. Then, the number of these tiny bits is assumed to be infinity.

A line has one dimension. A flat figure such as a square has two dimensions. The world that we encounter daily has three dimensions—length, breadth and height. Mathematicians have extended this concept and have defined infinite dimensional spaces. Such spaces do not really exist, but they are useful tools employed while studying quantum mechanics (which is a branch of Physics used to study particles of the size of atoms or even smaller), electromagnetic wave theory (which is the theory of light) and so on.

Another area where infinity finds use is fractals, which you will learn about in a later chapter. They are geometric figures, just like rectangles, circles and squares but have special properties. Many natural phenomena such as ferns, coastlines, even pine cones, are shaped like fractals.

Figure 2.1: Pine cones are shaped like fractals

Suppose you want to measure the coastline of Africa. If you measured it with a mile-long ruler, you would get a certain measurement. What would happen if you used a foot-long ruler the next day? You would get a much bigger answer. Using an inch-long ruler, you would be able to measure all the nooks and crannies in the coastline. As the length of the measuring instrument approached zero, your measurement would approach infinity!

In Physics, one comes across infinity in black holes. There are two opposing forces in a star. The heat present in the star causes it to expand and gravity pulls it together. These forces balance each other to keep the star in a stable state. When a large star runs out of fuel, it is unable to support its own weight. The force of gravity gets the upper hand and the star collapses. Eventually, the star gets even smaller than an atom. Imagine a star squashed into a space less than the volume of a tiny atom! Its density (which is the amount of matter present in a certain volume) is infinite for all practical purposes. The gravity of a black hole is so powerful that anything, including light that gets too close to it, gets pulled in.

Infinity can be found in art as well. When an artist tries to depict three-dimensional space on the two-dimensional space of a canvas, parallel lines (which should meet at infinity in three-dimensional space) are shown to meet at what are known as 'vanishing points' in the painting. This is in order to create an impression of depth. Artist M.C. Escher is known for using the concept of infinity in his work.

Infinity gives rise to many paradoxes. A paradox is an argument, which leads to seemingly absurd conclusions, due to a contradiction in logic. Hilbert's paradox of the Grand Hotel is one of them.

Figure 2.2: The concept of infinity in Escher's work

Normal hotels have a finite number of rooms. Imagine a hotel with an infinite number of rooms, which are all occupied. Now we have a new guest who wishes to be accommodated in this hotel. The person in Room No. 1 is shifted to Room No. 2, the person in Room No. 2 is shifted to Room No. 3 and so on. In a normal hotel, the last person would have nowhere to go. But in our hotel there is no last person! If we repeat this procedure, we should be able to make space for any number of guests.

Galileo's paradox is another example where the conclusion seems to go against common sense. You would have studied about sets. We have finite sets and infinite sets. With finite sets, a part is always smaller than the whole. But with infinite sets, a part can be just as large as the whole!

Let me explain. Some numbers are perfect squares (such as 1, 4, 9, 25 and so on), while others are not. Common sense tells us that there should be more natural numbers (i.e., numbers such as 1, 2, 3, 4, and so on) than numbers that are perfect squares. But this is not true. It is possible to pair natural numbers with their squares:

1	2	3	4
1	4	9	16

Both these lists go up to infinity. Thus, there are as many squares as there are natural numbers!

Georg Cantor was a famous German mathematician who studied infinity in great detail. He showed that there are various sizes of infinities—small infinities, large infinities, very large infinities and so on. I know that this is mind-boggling, but that is how things are in the world of abstract Mathematics. The largest infinity is known as the continuum. How large that would be is literally beyond comprehension!

Many mathematicians, who were Cantor's colleagues and contemporaries, did not accept his work at all, as Cantor's theory led to paradoxical results. Then they did not like the idea of infinite sets. They argued that since you can't list all the elements in an infinite set, anything you say about infinite sets is not really Mathematics. One of Cantor's main opponents was Leopold Kronecker, who used his influence to prevent much of Cantor's work from being published in his lifetime. However, Cantor's theories are widely accepted today, and his theory of infinite sets has created a strong foundation for all the branches of Mathematics.

The concept of infinity is difficult for the human mind to grasp. Everything we see has a finite size. It is impossible for us to imagine endless journeys or boundless spaces. It is because we cannot fully comprehend infinity that it remains fascinating, elusive and always out of reach.

3

THE MAGIC NUMBER PI

Pi or π is a Greek symbol and an important mathematical constant. You have studied circles previously. What is the diameter of a circle? If you draw a straight line that passes through the centre of a circle and begins and ends at two points on the circle, it is known as a diameter. The circumference of a circle is the length of the circle.

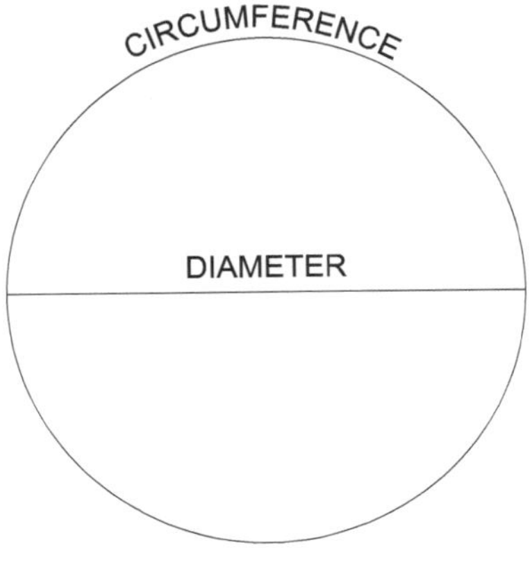

Figure 3.1: A circle

Now, if you divide the circumference by its diameter, the answer is π. The amazing thing is that it does not matter if your circle has a radius of 1 cm or thousands of kilometres. The ratio is *always* π.

Pi is an irrational number, which means that it can't be written as a fraction. This also means that it has an unending decimal number, where no pattern can be observed as such. The fraction 2/7, when written in the decimal form, is also unending. But after 6 digits, it repeats itself. It goes 0.285714285714285714...

The fraction 1/4 can be written as 0.25. That's all—just two digits. An irrational number such as pi does not end and does not repeat itself. The exact value of pi has remained a mystery since ancient times. Mathematicians who have continued trying to determine the value of pi can only come closer and closer to the truth, without actually getting there. Though the exact value will never be known, 3.14159 is a good approximation.

Pi is used extensively in geometry. For example, one needs the value of pi for evaluating areas of circles. You would have been introduced to solid objects such as the sphere, cone and cylinder. Pi is also used for calculating their volume and surface area.

Pi appears in—and is, in fact, quite essential to—numerous branches of Mathematics and Science—in statistics, the science of collecting and analysing large amounts of data; in thermodynamics, the study of heat and its effects; in mechanics, the science of building things; in electromagnetic theory, the combination of electricity and magnetism that is

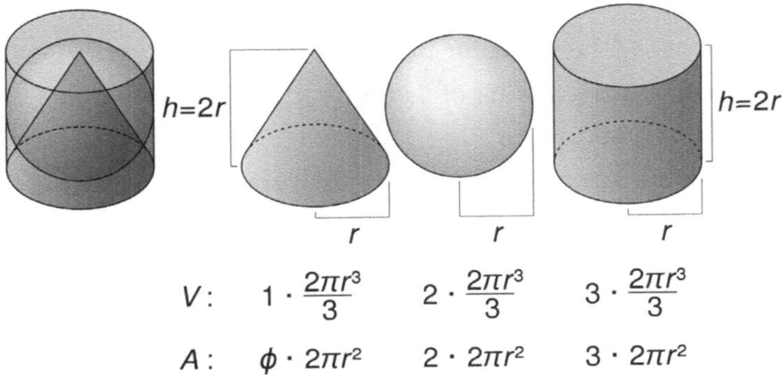

Figure 3.2: A cone, a sphere and a cylinder whose volumes and surface areas involve pi

the basis of all electronics and in cosmology, the study of the origin and evolution of the universe… The list is as endless as the value of pi itself.

Pi is so widely used that an entire day has been devoted to it! Pi Day is celebrated by Maths lovers in the US annually on 14 March, which is 3/14 in the month/day format that Americans use. 3/14 looks like 3.14, the simplest value of pi that is used. Get the logic? Some real fanatics even go to the extent of celebrating it exactly at 1:59 p.m.—as the exact value of pi is 3.14159!

Here are some more interesting titbits about pi:

- In an episode of the classic original *Star Trek* TV show of the 1960s, Dr Spock is able to cripple a very powerful and evil computer by commanding it to calculate pi to its last digit!

- Many movies have been made about pi. In *Pi: Faith in Chaos*, a man goes mad trying to make sense of pi.
- In the novel *Contact* by the cosmologist Carl Sagan (which was also made into a very successful film), humans study pi to gain deeper awareness of the universe!
- In 1992, Lars Erickson, then an engineering student at the University of Nebraska in the US, composed the pi symphony. He assigned the numbers 0–9 to ten musical notes, played them corresponding to the first thirty-two digits of pi and then built a symphony around this tune!
- 'I prefer pi' is a palindrome, which means that it reads the same forwards and backward.
- In 2010, two researchers in Japan wrote a special programme to calculate the value of pi and ran it on their computer. The computer worked for ninety days to produce a world record 5 trillion digits (5,000,000,000,000), before it gave up and crashed. But that record lasted only for three years. In 2013, a researcher at the Santa Clara University in the US calculated the value of pi to 8 quadrillion digits (that's 8 followed by 15 zeros). The latest record is by a woman, Emma Haruka, who calculated the value of pi upto 31 trillion digits.

People have been trying to calculate the value of pi for thousands of years. This ratio has been recognized for as long as we have written records. A ratio 3:1 appears in a Biblical verse as well. Ancient Egyptians calculated the area of a circle using the formula $(8d/9)^2$, where d is the diameter. This gives

3.1609 as the value of pi.

The first theoretical calculation of pi was performed by the great Greek mathematician Archimedes of Syracuse, who lived in the third century BC. He took a circle and drew a regular polygon each on the inside and outside. He began with a hexagon and went up to a polygon having 96 sides. He arrived at the following result:

$$223/71 < \pi < 22/7 \ (3.1408 < \pi < 3.1429)$$

After this, things were more or less quiet for about 2,000 years. Then, European mathematicians got into the act and started developing more and more complicated—but more accurate—ways to compute pi.

In the seventeenth century, Scotsman James Gregory and also the German mathematician Gottfried Wilhelm von Leibniz introduced the formula:

$$\pi/4 = 1 - 1/3 + 1/5 - 1/7 + \ldots$$

The only problem with this formula is that one needs 5 million terms to obtain the value of π correct to six or seven places of decimal!

People have devised even more difficult formulas to evaluate pi. These are used by computers today.

In 1777, French natural historian and philosopher Georges-Louis Leclerc Comte de Buffon devised a novel way to calculate π. Buffon approached the issue from a completely new angle, from the branch of Mathematics known as probability. The method consisted of drawing a grid of parallel lines and then randomly throwing needles and counting the number of times the needles fell across a line! The probability of a needle crossing a line is directly related to pi.

But why do people go to such lengths to find the value of π? I guess the reason is that pi is everywhere. Wherever there is a circle, we have π, yet it is not easy to measure it. You would not get a very accurate answer with a manual measurement of the circumference and diameter of a circle. As Mathematics progressed, so did our knowledge of π. And as computers became more efficient, we learnt to calculate it to a ridiculous degree of accuracy. Some people are mad enough to memorize thousands of digits of pi and then recite them on Pi Day!

Pi is so fascinating to mathematicians because it is one of those things that we will never be able to know fully, no matter how hard we try.

4

WHAT ARE RANDOM NUMBERS?

Look at the following series:

13, 26, 39, 52, 65, 78, 91...

1, 4, 9, 16, 25, 36, 49, 64...

1, 8, 9, 7, 28, 45, 0, 7, 89...

If you know your multiplication tables and the squares of natural numbers, you should be able to identify the first two series. What about the third series? Does it represent anything? Looking at any of the numbers, can you predict what the next number would be? The third series is a cluster of random numbers.

The numbers don't follow any pattern. You can't predict what will come next. These two qualities define randomness.

To a large extent, Mathematics is the science of patterns, and patterns do make Mathematics beautiful and a necessity to use in day to day life. We start with multiplication tables. 3, 6 and 9 come naturally and predictably. Later on, we study complex Mathematics and discover that there are more beautiful patterns to be found. However, randomness, which is also a part of Mathematics, is all about the lack of patterns.

Real life has randomness built into it. We have scientists trying to understand and control our physical environment. However, the physical world is uncertain, complex and changeable. The environment and even living creatures are constantly receiving and sending signals, which scientists then in turn study using instruments, numbers and graphs. When we speak, we send out sound waves as signals. Vibrations in the earth's crust and earthquakes produce seismological signals. Our biological processes also send out signals that can be measured. Most of these signals display some amount of randomness. All this makes us realize that randomness is a part of life, and if we are using Mathematics to understand the world, it should include randomness.

Tossing a coin gives a random result. People have used this to make important decisions or when they are trying to be fair—we often hear the phrase 'leaving things to chance'. During the Vietnam War in 1969, a lottery system was used to select the men who were to be recruited for military service. All the dates of the year were written down on small slips of paper and kept in a large bin and then picked up randomly. Young men between the ages of 19 and 25, born on those selected dates, were asked to serve in the American armed forces immediately.

However, this procedure was not random enough. The year-end dates had been thrown in last and were more likely to be drawn than those at the beginning of the year.

Another area where randomness comes in handy is while

writing codes. What are codes? If you want to send a message to a friend and you don't want anyone else to read it, you write it in code. You will be reading more about codes in a later chapter. The friend who receives your coded message will need something known as a 'key' to be able to read your message. Now, the key is a relatively small amount of information, but it has to be kept absolutely secret and no one should be able to guess what the key is. So it is useful to have a key that consists of random digits or alphabets to throw unwanted onlookers off the trail.

Coding or encryption is important because it allows you to securely protect information that you don't want anyone else to know, and so that this information remains between a select group of people. Individuals, businesses and even governments use it to protect secret information. Real life spies rely completely on codes.

Random numbers are also useful when we try to simulate real world phenomena. What is a simulation? In a simulation we create a model (sometimes a computer model) of a system for the watershed, a forest or even the organs of our body. This recreation is done in a controlled environment in order to understand the factors controlling the system and to predict the future behaviour of the system. A large number of random numbers is used to generate such a model. Such methods of computing are known as Monte Carlo methods. The name is derived from a region in Monaco where there are casinos. Gambling is also based on randomness.

Suppose you don't know how to find the area of a circle. You can place it inside a square and then fill the square with randomly thrown pebbles. The ratio of the number of pebbles inside the circle to the number of pebbles inside the square will give you an approximate ratio of the area of the circle to the area of the square. That was a simple Monte Carlo simulation technique.

The larger the circle and the number of pebbles, the more accurate are the results. Suppose we want to study the voting pattern in a particular state. We would take a small sample and study the pattern. This process is known as statistical sampling. And the best way to choose the sample would be by selecting candidates randomly.

How can we produce random numbers? One would think that if we randomly recite numbers, we would be able to generate them. However, humans have a tough time trying to act randomly. When people are asked to choose a number between 1 and 10, very few people would choose the numbers 10 or 1.

Traditional methods such as tossing a coin, spinning a roulette wheel or throwing a dice are still in use. Modern methods either employ physical phenomena or use a computer programme to generate random numbers.

There are two kinds of random numbers—true random numbers and pseudo (meaning fake) random numbers. True random numbers are generated by natural phenomena. For example, there is noise present in the atmosphere, and measuring it gives us random numbers. You have studied about electricity. When a current flows through a wire, we have electrons moving in a certain direction. In addition to this flow, the electrons also move randomly due to thermal (or heat) energy. This random movement causes a fluctuation in the electric current. This phenomenon is known as thermal noise. If we try to measure this noise, we will get random numbers.

You must be aware that the universe started with the Big Bang. There was a tremendous explosion that sent matter hurtling across space, as a result of which a huge amount of energy was produced. There is still some energy left over from the Big Bang. It is in the form of microwaves (the same kind of waves that heat up your food in a microwave oven). This is known as cosmic background radiation. Measurement of this radiation also gives us random numbers.

To generate pseudo-random numbers we use computer algorithms (which are a set of instructions given to a computer). We use a relatively short number known as the key or the seed. Based on that, a long string of random numbers can be generated. John von Neumann, who was a great mathematician, came up with the following method for generating random numbers.

$X_0 = 35385906$ was the number he started with.

Then he squared the number.

$X_0^2 = 1252162343440836$.

After that he took the eight digits in the middle.

$X_1 = 16234344$.

Then he squared this number and repeated the process.

This repetitive process is known as an algorithm and it generates a large number of random numbers. But the pattern repeats itself after a while. People are trying to devise new and better ways to generate random numbers.

There are other areas of Mathematics where randomness is used. A random walk is a mathematical concept which describes a path that consists of a series of random steps. The path traced by a molecule in a liquid or a gas can be modelled as a random walk. The financial position of a gambler can also be described as a random walk!

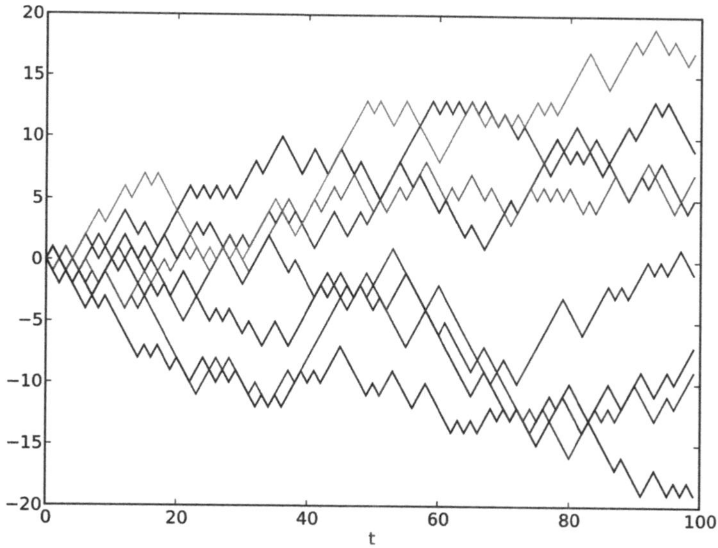

Figure 4.1: A random walk

While taking any measurement using an instrument, there is always an unpredictable amount of error present. This is known as random error. It affects the accuracy of measurements. However, random error can be greatly reduced by taking a large number of readings and finding the average value.

Random words are used to think creatively. Suppose you

are in the middle of a creative project—for example, an essay. When you get stuck, you can think of a random word and then look for words associated with it. This may help you to think in a different manner.

People also behave unpredictably and randomly. With so much randomness all around, Mathematics would be incomplete without the consideration of random numbers.

5

FLIGHTS OF FANTASY: IMAGINARY NUMBERS

We have different names for different kinds of numbers. 1, 2, 3, 4, etc., are called natural numbers. We use them for counting objects. Zero signifies nothingness and is useful when we are trying to describe large numbers. We also have negative numbers, which are helpful when we are talking about owing people something.

Fractions and decimals allow us to divide things equally between people. Suppose you have 3 pieces of chocolate and there are 5 people, each person would get 3/5 of a chocolate. Then, we have irrational numbers. These are numbers that cannot be written as a fraction or a decimal. π and $\sqrt{2}$ are examples of irrational numbers.

All these numbers are known as real numbers because they have some relevance to the real world. Have you seen the number line? It is a straight line that extends from minus infinity to plus infinity. Each real number represents a point on the number line and signifies a certain length.

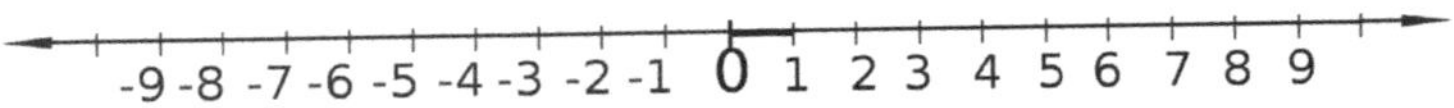

Figure 5.1: The number line

There is another class of numbers known as imaginary numbers. These numbers have no equivalent in the real world at all—mathematicians just created them out of thin air!

Let us see what exactly imaginary numbers are. When you square a positive integer like 2, you get another positive integer, 4. When you square –2, you get 4 again. So the number 4 has two square roots: +2 and –2. What about –4? It doesn't seem to have a square root at all. Well, mathematicians have invented a square root for –4. It is an imaginary number and equal to 2i, where 'i' is the square root of –1.

The imaginary number 'i' was born out of necessity. In AD 50, Heron of Alexandria, who was a brilliant mathematician, was trying to find the volume of an impossible section of a pyramid. The answer was the square root of a negative number.

He was forced to give up because there was no concept of imaginary numbers at that time. The same problem showed up when people had to find the solutions of equations such as $x^2 + 1 = 0$. It was in AD 1545 that Girolamo Cardano, who was an Italian mathematician and a polymath (a brilliant person with interest in several fields of study) who lived in the sixteenth century, solved an equation using imaginary numbers. But he was not too pleased with the idea. He called the idea of imaginary numbers a 'mental torture'.

However, we can understand the concept of imaginary numbers better when we compare them to negative numbers. Negative numbers were considered absurd till the seventeenth century. Imaginary numbers are considered absurd even today. A negative number is strange because it is less than nothing, and, therefore, hypothetical. An imaginary number is strange because it is the square root of a number which is less than nothing.

What is the intuitive meaning of negative numbers? It means opposite. When we go backwards on the number line, we get negative numbers. The intuitive meaning of imaginary numbers is rotation. In order to understand this, we have to understand complex numbers and the complex plane. A complex number is the sum of a real number and an imaginary number. It can be written as:

$$z = (a + ib)$$

Here 'a' is the real part, and 'b' is the imaginary part. Each complex number can be represented by a point on the complex plane. The complex plane is an abstract idea and a useful concept. You would locate a point on the complex plane the same way as you would locate a point on a graph sheet. You go 'a' cm to the right and 'b' cm upwards. The complex plane helps us visualize complex numbers. You could think of complex numbers as two-dimensional numbers.

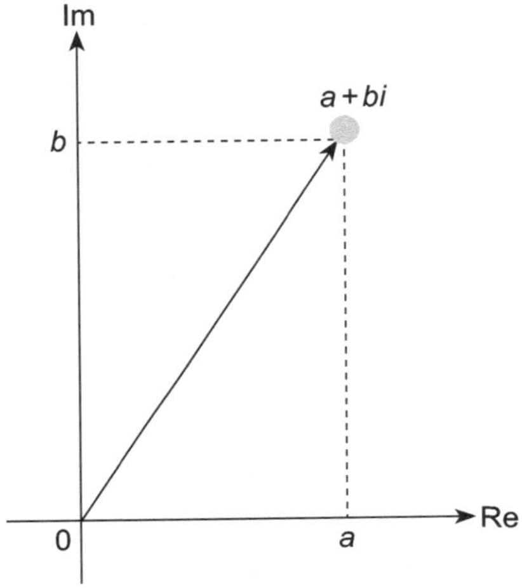

Figure 5.2: A complex plane

Now, the number 1 has a unit length and lies on the real axis. The imaginary number 'i' also has unit length, but lies on the imaginary axis. Thus, 1 multiplied by 'i' indicates a rotation on the complex plane by 90^0. When we multiply 'i' by itself, we get −1. This is another rotation by 90^0 in the counterclockwise direction. Generalizing, i^n signifies n rotations of 90^0 on the complex plane.

Here are two examples to illustrate this:

When we take the real number 4 and multiply it by −1, we get the following pattern:

4, −4, 4, −4, 4, −4…

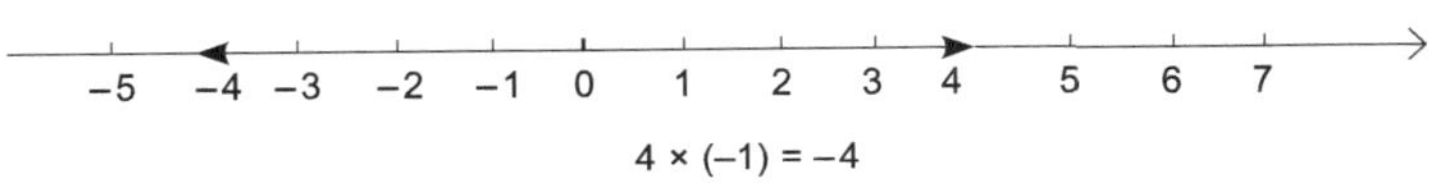

Figure 5.3: Multiplication by −1 on the number line

Thus, we see that multiplication by –1 means moving in the opposite direction in the number line.

When we start with 3 and multiply by i, we get the following pattern:

3, 3i, –3, –3i, 3,…

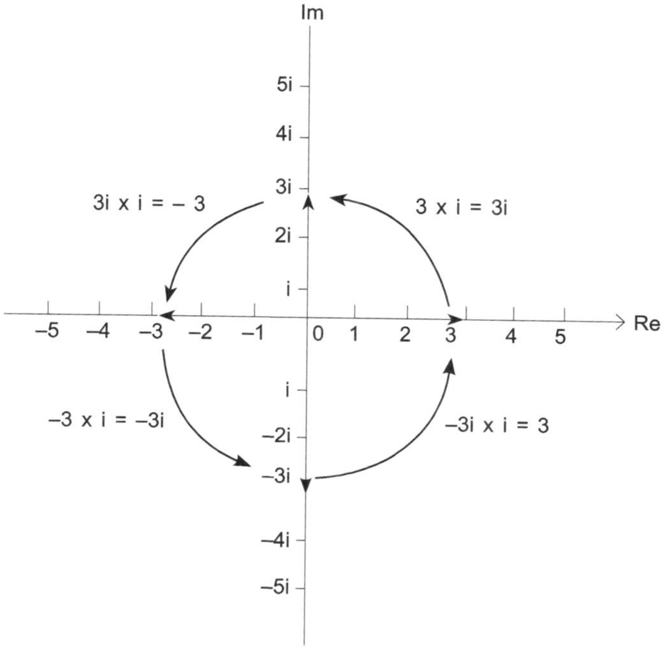

Figure 5.4: Multiplication by i

Thus, multiplication by 'i' indicates a rotation of 90^0 on the complex plane.

Let's take another example. The complex number (1 + i) makes 45^0 with the x axis.

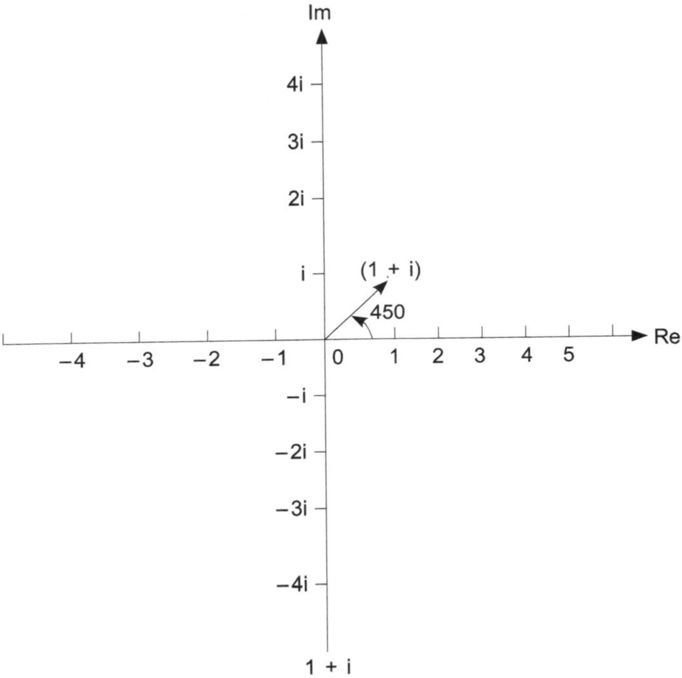

Figure 5.5: Multiplying a complex number by (1 + i)

If we multiply a complex number by (1 + i), it will produce a rotation by 45°. Let us multiply and see what happens:

$$(3 + 4i)(1 + i)$$
$$= 3 + 3i + 4i + 4i^2$$
$$= 3 + 7i - 4$$
$$= -1 + 7i$$

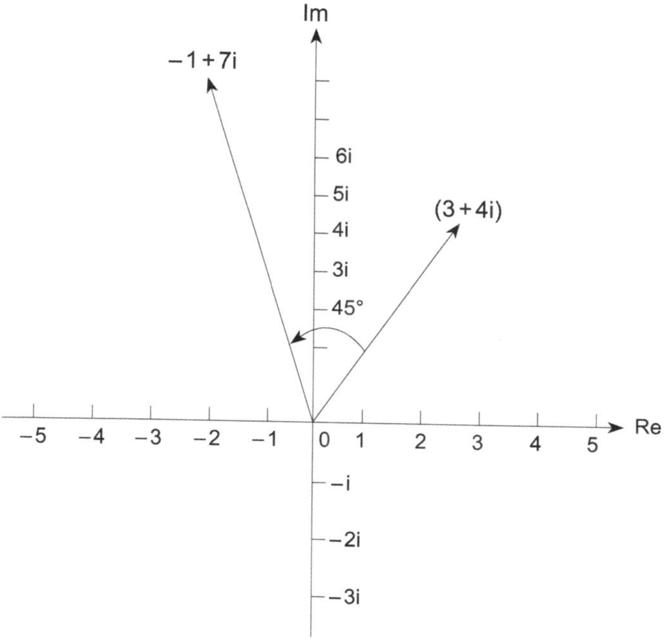

Figure 5.6: Multiplying by (1 + i) gives a rotation of 45°

Let us try and understand what this physically indicates. If we have a direction 3 units east and 4 units north and we rotate it by 45^0, we will obtain a direction 1 unit west and 7 units north. Thus, multiplication by complex numbers makes rotation very easy.

Complex numbers find application in a wide variety of fields. The algebra of complex numbers greatly simplifies problems in Physics and Mathematics. Engineers use complex numbers in analysing stresses and strains on beams. (Stress gives you an idea of the weight a beam is carrying and strain gives a measure of the extension caused by the weight).

Complex numbers are used in studying the flow of fluids around obstacles, such as the flow around a pipe. Transmitting radio waves, talking on the cell phone and listening to the radio would not have been possible without complex number analysis. Imaginary and complex numbers are also useful while summing up an infinite series and solving algebraic equations.

Although imaginary and complex numbers are based on abstract ideas and are hard to grasp, their algebra is quite easy and they do make life simpler for mathematicians, physicists and engineers. Philip Pullman, the British writer, has likened imaginary numbers to Adam and Eve. They may or may not have existed, but if you include them in your equations, you can calculate all manners of things, which could not have been imagined without them!

6

HOW MANY DIMENSIONS ARE THERE?

Think of a point. It has no dimension at all. It cannot have a length, breadth or height. A line, which is a set of an infinite number of points, has only one dimension. You can move along it in just one direction. A plane figure such as a square is two-dimensional. A plane is obtained by dragging a line perpendicular to itself.

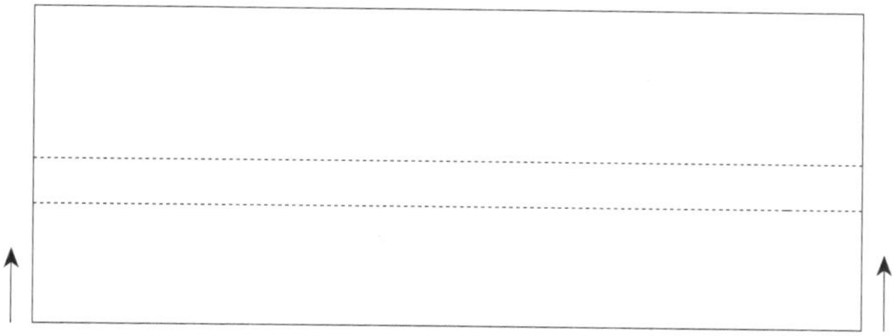

Figure 6.1: A line dragged perpendicular to itself forms a plane

In a plane, you can move in two independent directions—you can move up and down or right and left. A cube is three-dimensional. If you raise a square through a distance equal to its height, it would have traversed through a cube.

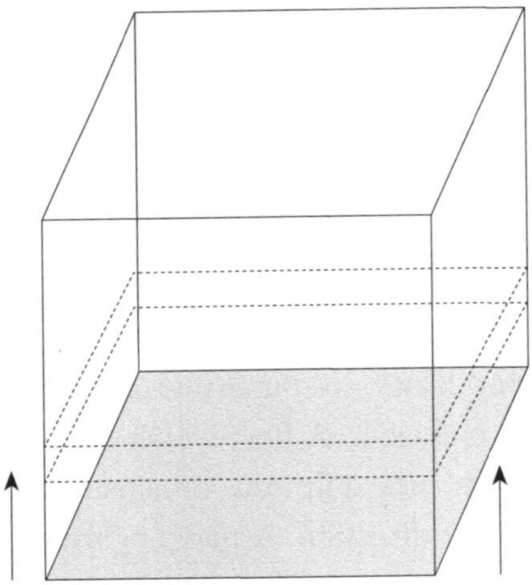

Figure 6.2: A square dragged perpendicular to itself forms a cube

The length, breadth and height of a cube give it solidity. One can study a two-dimensional plane or a three-dimensional space using a branch of Mathematics known as co-ordinate geometry. You have learned how to draw graphs. Each point on the graph has an x co-ordinate and a y co-ordinate. When we are talking about the third dimension, we would need the z-axis in addition to the x and y axes.

Can we extend the concept of three-dimensional space to four-dimensional space? We live in a three-dimensional world where the objects we see around us and interact with are three-dimensional. However, mathematicians have taken this a step further by creating the concept of 4D space using the same rules that they applied for 3D coordinate geometry. Except that now there are four coordinates.

Now, what does this mean to think of something as four-dimensional?

Think of a circle. It is a set of points on a plane having the same distance from its centre. Its boundary is a line known as the circumference. A sphere is a 3D version of a circle. It is a set of points in three-dimensional space, all of which are equidistant from the centre. Its boundary is a curved surface. When we get into 4D, we obtain a hypersphere. A hypersphere is a set of points in four-dimensional space having the same distance from a given point.

In order to get an intuitive idea of the hypersphere, let's try to use our imagination a bit. Suppose a spherical ball passes through a plane surface. We would first see a point, then a growing circle, followed by a diminishing circle and then finally a point again. Now if a hypersphere was to pass through 3D space, we would first see a point, then a growing sphere, followed by a diminishing sphere and then a final point again!

A cube is a three-dimensional version of a square, and a tesseract is a cube in a four-dimensional space. How would you obtain a tesseract? A point, when pulled along in a particular direction, gives us a line. A line, when pulled along a direction perpendicular to itself, gives us a square. A square, when dragged along a direction perpendicular to itself, gives us a cube. To obtain a tesseract we would have to pull the corners of a cube in a direction perpendicular to all the three axes—x, y and z. This is almost impossible for us three-dimensional creatures to imagine. A tesseract is often depicted by placing a smaller cube inside a larger one.

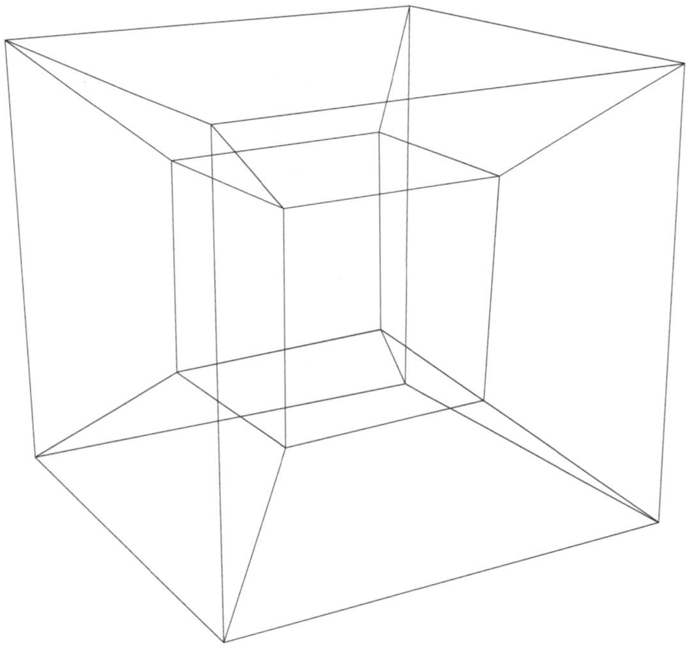

Figure 6.3: A tesseract is depicted by placing a smaller cube inside a larger one

If we had 4D creatures, what would they be able to do?

Imagine a two-dimensional creature walking on a flat surface. As it approached a square, it would just see a straight line, which is the boundary of a square. We are able to see the entire square—the boundary and the inside. In fact, we see everything in 2D and are able to form a 3D picture with the reasoning faculty of our brains. Extending this analogy, a four-dimensional being would be able to see the world in 3D. It would be able to see all six sides of an opaque box as well as what is inside the box!

Mathematically and intuitively, we now understand dimensions. But how do dimensions apply to the physical world? The universe exists in three-dimensional space. You can think of the fourth dimension as time. So in four dimensions, we can capture the history of the universe—which is something like watching a 3D film.

Let us take it one step further. If we have five dimensions, we can have every possible history of the universe—for example, a universe where Hitler won World War II, and a universe where dinosaurs never went extinct. If we take the sixth dimension into account, we can have all possible values of the physical constants. For example, the gravitational constant and the speed of light would have all sorts of values. The structures of these universes would be vastly different.

Extra dimensions come in very handy in Physics where an understanding of forces is very important. A force is a push or a pull on an object, which can cause an object to start or stop moving. It can also change the direction of a moving object or the shape of a stationary one. There are four fundamental force in nature—gravity, electromagnetic forces, the weak nuclear force and the strong nuclear force. Interaction between particles is caused by one or more of these forces. Gravity is what keeps the planets moving around the sun and the moon around the earth. We can't imagine life without electricity. All our electrical gadgets work because of electromagnetic forces. The strong nuclear force is what keeps the nucleus of an atom together. And the weak nuclear force comes into play during radioactivity, which is the spontaneous emission of radiation from unstable nuclei.

What physicists have been trying to do is develop a theory that unifies all the four forces mentioned above. String theory is one of the most popular of the various theories that attempt the unification of all four forces. According to this theory, the fundamental building blocks of the universe are string-like. While they possess only one dimension, they vibrate in multiple dimensions, and depending on how they vibrate, they appear as light, matter or gravity. These strings don't exist physically. They are mathematical concepts that explain the working of the universe.

Superstring theories originate from a more fundamental eleven-dimensional theory known as M-theory. Many physicists believe that M-theory could explain everything in the universe. We only have to believe that all these extra dimensions exist!

We also come across extra dimensions in Einstein's special theory of relativity. Everyone has heard of Albert Einstein, who was one of the smartest scientists the world has seen. His theory of special relativity changed the way we look at the world and defied our common sense notions of space and time. The equations of the theory of special relativity can be applied very well in a four-dimensional space known as Minkowski space.

Science fiction writers have created parallel universes with extra dimensions. People from all spheres of life have not confined themselves to just three dimensions. With the help of your imagination you can create as many dimensions as you want anytime!

7

CREATING CHAOS IS SO EASY!

To a layperson, the term 'chaos' suggests complete disorder and confusion. But in Mathematics, 'chaos' denotes something else. Chaos theory is a branch of Mathematics that studies complex systems. Let's take a few examples to understand what chaotic behaviour is.

Take a simple pendulum. It oscillates to and fro quite predictably till the oscillations die down. Now if you attach a second pendulum to the tip of the first, you will find that the second pendulum oscillates in a completely haphazard manner.

You may be aware that Sir Isaac Newton, the great physicist, discovered the law of gravity and the three laws of motion. He used these to demonstrate how the solar system worked. However, there was a slight problem. The theory worked beautifully only if there were just two bodies. The eminent French mathematician Henri Poincaré, who lived during the end of the nineteenth century, tried to introduce a third body and solve the same equations. He was unable to get any solution. He also found that small changes in the initial positions of the three bodies would lead to large changes in the paths traced out by the bodies. These are ideas that lie at the heart of Chaos theory.

Poincaré was the first person to study chaotic systems. However, mathematicians and physicists nearly forgot about Chaos theory for about a hundred years. In 1960, Edward Lorenz, a mathematician and a meteorologist, created a weather model on his computer at MIT. His weather model used a complex set of formulas that used lots and lots of numbers. With the help of these numbers Lorentz was able to 'create' clouds, winds, heat and cold! His colleagues and students were deeply impressed by the weather model because it never repeated a sequence, and the results were quite like the actual weather. Some people even hoped that Lorenz would build an ultimate weather predictor one day.

On a certain winter day in 1961, Lorenz decided to re-examine the output from his machine. Instead of running the entire programme, he started from somewhere in the middle and let his computer perform the calculations all over again. What he found was quite unexpected. The data obtained from the re-run should have been identical to the data from the first run. However, while the data did match initially, after a while, the results obtained were dramatically different. Lorenz soon realized that this had happened because he had rounded off the input data to three decimal places instead of retaining all six decimal places of data.

Lorenz realized that long-term weather forecasting was bound to fail. Small changes in the initial conditions would lead to large changes in the result. This is referred to as 'the butterfly effect', which literally means that a butterfly flapping its wings in Brazil could cause a hurricane in South China Sea!

Chaos theory is used to study complex systems. In classical Physics, when we are dealing with topics such as Newton's laws of motion, electricity and magnetism, we get predictable results. If we know the starting point, we will know what the end result will be. However, Chaos theory deals with systems that are impossible to predict or control. And small changes in the initial conditions lead to large changes in the final result.

Complex systems contain so many elements that move in so many different ways that we need computers to calculate all possible outcomes. Chaos theory emerged only in the second half of the twentieth century because there were no computers before that. Another reason why Chaos theory was born so recently is that quantum physics (which is a branch of Physics which deals with particles smaller than the atom) was introduced to the world in the first half of the twentieth century. Before quantum physics was born, people believed in cause and effect. They believed that a certain action would always produce the same result. They also believed that if we could catch and measure the speed of every particle in the universe, we would be able to predict everything using mathematical equations.

Quantum physics showed us that there would always be a certain amount of uncertainty in the initial conditions of actions which would lead to unpredictable results. This is the basis of Chaos theory.

Complex systems are so chaotic because of the number of

variables that it is often difficult for us to recognize patterns in them. However, with the help of certain techniques, we can use a point on a graph to represent the whole system. Then we can let the computer run its programme and we can watch how this point traces a pattern.

The first Chaos theorists discovered that complex systems also run through some kind of a cycle and the system tries to achieve an equilibrium of some sort. The equilibrium state is called an attractor. Imagine a school with a thousand students. They have a good student-teacher ratio. There are a hundred teachers. Forty teach Humanities, forty teach Science and the rest teach the Creative Arts. This is some sort of equilibrium (meaning stable state). Everyone is content with the situation. Suppose a hundred students are added. They add some more teachers in the same proportion and equilibrium is maintained again. This new equilibrium is called an attractor.

Now imagine that ten Science teachers leave and join some other school. The kids are left without a Science teacher for six months. By the time the new teachers join, the students have got used to the free time and have lost interest in Science. The new teachers are unable to maintain discipline in class. All of them feel demotivated and one by one, they leave. Again, there is a gap of a couple of months and the same situation keeps repeating itself. This unpleasant situation is also an equilibrium of some kind, but a dynamic one, meaning that things keep moving back and forth between states. A dynamic equilibrium is called a strange attractor.

The difference between an attractor and a strange attractor is that an attractor indicates an equilibrium state to which a system finally settles down, whereas a strange attractor represents a sequence of events that keeps repeating itself, maybe approximately, without ever settling down.

The discovery of attractors was very exciting and illuminating, but the most wonderful thing that Chaos theorists discovered was something known as self-similarity. Self-similarity helped us gain a better understanding of the mechanisms that shape the cosmos, the world and even ourselves!

What is self-similarity? It is a fundamental principle of nature that allows the building blocks of nature to copy their own shape over and over again in the structure that they create. This can be seen, for example, in the way trees branch out.

Self-similarity is a concept that has far-reaching consequences. It is observed in all of nature. Many believe that self-similarity is one of the fundamental principles that shape our world. Self-similarity has been observed in all fields of research—Physics, Biology, even Psychology and Sociology.

Where can we apply Chaos theory? The earth's weather system can be described using Chaos theory, along with something as basic as the motion of water molecules as observed while boiling water. It can also be applied when we are studying the migratory patterns of birds, or even the spread of vegetation across continents. Even in films, when we observe a lot of movement and objects that are also in motion, we are actually watching chaos-based graphics.

The stock market is a good example of a chaotic system. When the price of a stock rises or falls, people scramble to buy or sell that stock. This has a direct impact on the stock price, and the price rises or falls chaotically.

Chaos makes us view the world in a different way. The traditional sciences deal with phenomena such as gravity, electricity or chemical reactions, which are predictable. Chaos theory deals with phenomena, which involve several factors and not simple cause and effect, such as the weather, stock prices or our brain states, which are impossible to predict or control. However, one can use fractal mathematics to study them. Chaos theory was derived from weather prediction; yet today it has a large range of applications. By understanding that our ecosystems, social systems and economic systems are actually interconnected, we can make better decisions for ourselves and society at large by avoiding actions that might seem beneficial right now, but would be harmful for all of us in the long run.

8

THE ENDLESS WORLD OF FRACTALS

Extremely simple, yet incredibly complex, fractals are endlessly fascinating. They are the meeting point of Mathematics and Art, computation and the natural world. Fractals have always been around, but mathematicians started taking an interest in them in the 1970s. A fractal is a geometric pattern that is repeated endlessly at ever smaller (or larger) scales to produce irregular shapes that cannot be represented by geometry as we know it. They are found in all sorts of scales, from a spiral galaxy thousands of light years away, to a perfectly symmetrical snowflake.

Benoit Mandelbrot, the twentieth century mathematician, coined the word 'fractal' in 1975, meaning irregular or fragmented. Not too long before that, Lewis Fry Richardson discovered this phenomenon while studying the length of the English coastline. He realized that the measurement would be greatly dependent on the instrument that was used to measure it. If he used a measuring instrument that was a mile long, he would obtain a certain answer, but a foot-long ruler would yield a much larger result. And if he took an inch-long apparatus to measure the length of the coastline, he would be taking into account all the nooks and crannies, and obtain a much larger answer. This endless increase in the length of the coastline was happening despite the fact that the area of the land that it encompassed remained fixed.

The beauty of a fractal is that although the pattern is quite elaborate, it starts with something simple.

As an example, take a triangle as a starting point. Shrink it to half its size so that you obtain another triangle. Now take four triangles congruent to the smaller triangle and insert them in the larger triangle. One can go on repeating this procedure forever. The figure that we obtain is a fractal. This is known as the Sierpinski triangle.

Figure 8.1: The Sierpinski triangle

Our starting point need not have been a triangle. We could have started with almost any shape or figure and with repetition, obtained a fractal.

The really cool thing about fractals is that they have a fractional dimension. A fractal's dimension may be 1.2. Let us try to understand what this means. Suppose you have a 1 metre long line, which is one-dimensional. If you tried to measure it with a 50 cm scale, you would get a measurement of 2. If you have a square of side 1 m (which is two-dimensional) and you try to measure it with a square of side 50 cm, you would get a measurement of 4. If you tried doing the same thing with a fractal you would not obtain 1 or 4. You would get a value in between. This tells us that a fractal is not as simple as a one-dimensional object, but not as complex as a two-dimensional one. It has a dimension between 1 and 2.

Fractals are found everywhere in nature. The laws of the natural world are helpful in the creation of fractals. The growth of pineapples seem to be governed by these laws. So are ice crystals. Fractal shapes show up in river deltas and in the veins of our bodies. They are efficient as they allow plants to maximize their exposure to sunlight, and our circulatory system to efficiently transport oxygen to all parts of the body.

Figure 8.2: Ice crystals

Our brains are filled with fractals as well! The human brain has 100 billion neurons or brain cells, among which there are about 100 trillion synapses (or connections). There are two kinds of branched projections from a brain cell, called axons and dendrites. The axons of a neuron reach out to make synaptic connections with the dendrites of other neurons in order to pass electrical stimuli from one neuron to another. It is because axons and dendrites have fractal structures that they are able to communicate with so many cells.

A broccoli is shaped like a fractal. So are pine cone seeds. Many plant leaves grow out as fractals. They can also be seen in the formation of ice crystals, trees, rivers, leaves, raindrops and even air bubbles.

Figure 8.3: A tree, an example of a fractal

Chaos and fractals are described using equations, as when we plot a graph for chaotic behaviour, we tend to obtain fractal-like patterns.

Apart from being beautiful and mesmerizing, people study the phenomenon because they have many applications in science. The reason is that they often describe the real world better than traditional Mathematics and Physics.

Fractals could potentially revolutionize the way we perceive the universe. Cosmologists usually assume that matter is distributed in a uniform manner across space. But observation shows that this is not true. The universe does seem to have a fractal-like structure on a small scale. No one really knows the big picture, but a group of scientists claim that the structure of the universe is fractal at all scales! In order to prove this theory, we would need a lot of experimental evidence.

Many things in nature—trees, cloud formations, the changing weather, the movement of water currents or animal migration patterns have a fractal nature and can be described by Chaos theory. With fractal geometry, we can visually model much of what we witness in nature such as soil erosion and the patterns of earthquakes.

Computer images are often compressed so that more images can be stored in a given amount of memory space. For this purpose, fractal image compression is employed. The basis of this technology is that the real world is so well described by fractal geometry, and so, using this technology, we are able to compress images to a greater extent.

9

THE PYRAMIDS ARE REALLY STRANGE!

The pyramids, as you may already know, are the grand stone tombs of the kings of ancient Egypt—the pharaohs. They are one of the seven wonders of the ancient world, and the only ones that still exist. More than 15 million tourists from all over the planet visit the pyramids every year.

The Egyptians believed in mummifying their pharaohs so that they could live on forever. The tomb was designed to protect the pharaoh's body. Egyptians placed treasures inside the burial chamber to make sure that the pharaoh's afterlife was also good. They also placed paintings to show the gods what a good man and king the pharaoh was. Mummified pets and even servants of the pharaoh would be put inside so that they could be of service in the pharaoh's afterlife.

Most of the pyramids were built on the western side of the Nile, just a bit into the desert. The dryness of the desert helped in preserving the pharaoh's body. The Great Pyramid of

Cheops at Giza is the largest and most famous of the pyramids, which was built for the pharaoh Khufu. Its height is more than 140 metres and it is estimated that it took the Egyptians 20 years to build.

The pyramids remain one of the world's greatest historical mysteries. And it's not only archaeologists and historians who are fascinated with them. Mathematicians and astronomers are, too! This is because inside the Great Pyramid, there are some of the strangest mathematical puzzles ever. Are these just coincidences, or did the ancient Egyptians know far more about the geometry of the universe than we give them credit for? Or, as some nutty people believe, was the Great Pyramid built by a super-intelligent race from another planet that decided to leave unending puzzles for us earthlings to ponder on?

The Orion Mystery

Think about this. The pattern the pyramids form on the ground is very similar to the pattern formed by the stars of the Orion constellation.

The Orion Mystery: Unlocking the Secrets of the Pyramids written by Robert Bauval and Adrian Gilbert caused a huge uproar in 1994. The authors of the book claimed to have finally unearthed the secrets of the pyramids. According to the book, the Egyptian world of the dead was located in the sky. They believed that their dead kings would become stars of Orion. Thus, the pyramids were placed in a pattern similar to that of the Orion constellation. Is the similarity to the Orion constellation pure coincidence or did the Egyptians know a lot of astronomy and geometry?

More Coincidences

You would have studied the globe and must be aware of latitudes and longitudes. The latitude and longitude lines that intersect at the pyramids cross more of the land's surface than any other line. The pyramids are thus situated at the exact centre of the landmass of the Earth. The original perimeter of the pyramid is exactly half a minute of latitude at the equator and the average height of land above sea level on the Earth happens to be the height of the pyramid. The curvature designed into the faces of the pyramid exactly matches the radius of the Earth. Are these mere coincidences or did the ancient Egyptians have a superior knowledge of the Earth's dimensions? Since the Earth has enough land area to provide 3 billion possible building sites for the pyramid, the odds of it having been built where it is are 1 in 3 billion.

Accuracy of Alignment

The sides of the pyramid run in the north-south and east-west directions. The accuracy of this alignment is extraordinary. There is an average variation of less than 0.06 per cent. Did the Egyptians use the stars or the rising sun as their references? We shall never know. The foundation of the pyramid is amazingly level. No corner is more than half an inch higher or lower than the others and the four corners are almost perfect right angles. What were the tools used by the builders to obtain this level of accuracy? Surely they didn't have all the electronic measurement technologies that we have now! So what did they use?

The orientation of the Great Pyramid is within 3/4 inch of True North—the exact direction along the Earth's surface towards the North Pole. How did the Egyptians, who live so very far away from the North Pole, know that?

The Pyramids and Pi

You have read about pi. If one divides the perimeter of the Egyptian pyramid by its height, one obtains a close approximation to 2π. Is this again a coincidence or did the pyramid builders have knowledge of π? The earliest written approximations of π were found in Egypt and Babylon.

The Pyramids and the Golden Ratio

In Mathematics and Art, two quantities are said to be in the golden ratio if the ratio of the sum of the quantities to the larger quantity is equal to the ratio of the larger quantity to the smaller one. It is represented by the Greek letter phi (Ø) and has an approximate value of 1.618. Since the time of the ancient Greeks, artists and architects have used this ratio in their work, believing it to be aesthetically pleasing. The ratio of the surface area of the Great Pyramid to its base is Ǿ, the golden ratio, which is another mysterious feature.

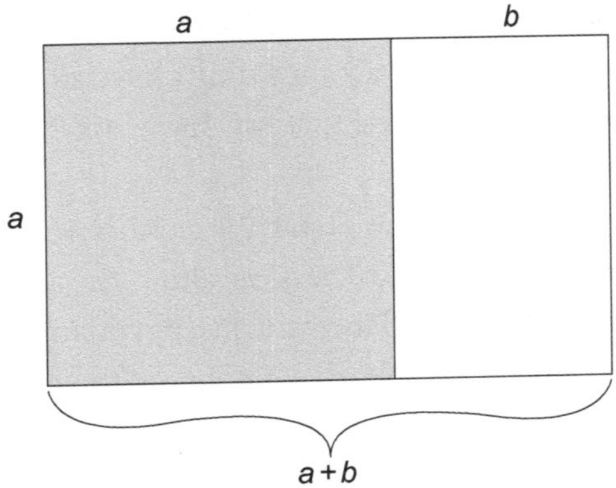

Figure 9.1: The Golden Ratio: the ratio of the length to breadth of the bigger rectangle is the same as that of the smaller rectangle. $\left(\frac{a+b}{a} = \frac{a}{b}\right)$

Squaring the Circle

There was a problem posed by ancient mathematicians. Using a straightedge and compass only, one had to draw a square with the same perimeter (or length of outer boundary) as a given circle. Many people spent a lot of time trying to solve the problem. It was proven in the late nineteenth century that the task was impossible. However, if you took the height of the Great Pyramid as the radius and drew a circle, it would have the same perimeter as the base of the pyramid! Were the ancient Egyptians aware of this particular mathematical problem and was the Great Pyramid their answer?

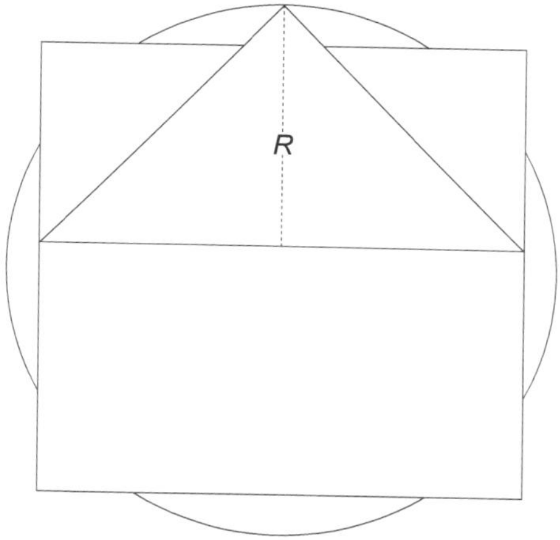

Squaring the circle and the great pyramid:
Radius of the circle = height of the pyramid

Figure 9.2: The square base of the pyramid has the same perimeter as the circle

The Pyramids and the Nautical Mile

The nautical mile is a unit of length related to the length of longitudes. It is used by sea and air navigators. The perimeter of the square base of the pyramid happens to be half a nautical mile. Is this another big coincidence or did the Egyptians know the dimensions of the earth?

The Speed of Light and the Pyramids

The speed of light is 299,792,458 m/s, and this is a universal constant. Now the precise latitude of the centre of the Grand

Gallery in the Pyramids of Giza is 29.9792458^0 N! Isn't that amazing?

Dimensions of the Solar System

The Earth's closest approach to the sun is 146×10^7 (i.e., 146 million) km. When this figure is converted into royal cubits (which is an ancient unit of measurement defined as the distance between the elbow and the tip of the middle finger), the answer is 280×10^9 royal cubits. The height of the great pyramid is 280 royal cubits. There is a granite coffer inside the pyramid. If you take twice its perimeter and multiply it by 10^8, you will obtain the sun's mean radius. The coincidences are far too many to be completely ignored.

More Mathematics

Have you heard of the Fibonacci series? The first few terms are given below:

0, 1, 1, 2, 3, 5, 8…

What we see is that the sum of any two consecutive numbers in the series is equal to the next number. The Fibonacci and other summation series are present in nature. The proportions of the structure of the pyramids show us that the ancient Egyptians who built them must have been aware of this series.

The pyramids continue to remain a mystery. There are people who have spent their entire lives trying to unlock the secrets of these amazing structures and the minds of the people who built them. But, in the end, all we can do is just wonder.

10

WHY ACHILLES CAN'T EVER OUTRUN THE TORTOISE—THE PARADOXES OF MATHEMATICS

Let me tell you a story. The tortoise once challenged Achilles, a great hero of Greek mythology, to a race. He said, 'Give me a headstart and I will win the race.' Achilles was amused by this because he was a mighty warrior and swift-footed as well, whereas the tortoise was slow and heavy.

'How big a headstart would you need?' he asked the tortoise just to humour him.

'Ten metres,' answered the tortoise.

'You know that I can cover ten metres in a very short time,' said Achilles.

'But in that short time how much distance would I have covered?' responded the tortoise.

'Maybe a metre,' answered Achilles thoughtfully.

'And you would be able to cover one metre very quickly?' asked the tortoise.

'Very quickly indeed,' answered Achilles.

'But by that time I would have moved further away,' said the tortoise. Now Achilles understood where the tortoise's argument was leading to. By his logic, Achilles would never be able to outrun the tortoise.

This was one of Zeno's paradoxes. Zeno of Elea (490–430 BC) devised a set of philosophical problems to show that motion is nothing but an illusion. The catch in the tortoise's argument is that an infinite number of time periods when added together can give a finite answer. For example:

$$1 + 1/2 + 1/4 + 1/8 + \dots = 2$$

Paradoxes are interesting because they challenge the logic that we utilize intuitively. They challenge our basic assumptions and make us think very hard. Puzzling over many of the known paradoxes has led to significant advances in Science, Philosophy and Mathematics. We have already discussed Hotel Infinity and Galileo's paradox. Let us explore some more paradoxes and see how challenging they are. Here's Cantor's paradox:

A set in Mathematics is a collection of well-defined objects. Its subset is a set contained within it. The power set is an important concept. The power set of a given set is the set of all the subsets within it. By definition, the power set is bigger than the set itself. Now imagine the Set of All Sets. It should be the biggest of all sets. But this set would also have a power set, which is bigger. So the set of all sets is a paradox.

Cantor did a lot of work on infinite sets and his work led to many paradoxes. He showed that there is a hierarchy of infinite sets. The set of all natural numbers is an infinite set. So is the set of all even numbers. They have the same size because we can show a one-to-one correspondence between the two sets. However, the set of real numbers is also infinite, but 'bigger' than the set of natural numbers. The set of all sets, which would be the biggest set, does not exist. Given a set, there would always be a set bigger than it.

Now let's examine Curry's paradox. It's a self-referential statement, which means that the statement refers to itself. Here's what it says:

'If this statement is true, Germany borders China.'

Let's try and understand why this is a paradox. Let us first assume that the statement is true. Then Germany should border China. That, we know, is false. Now let's assume that the statement is false. In that case, China need not border Germany. But that is the truth. That's why the statement is a paradox. 'I am a liar' is also a similar paradox.

Then, we have something known as the potato paradox. It is a term used to describe a mathematical calculation, which gives a counter-intuitive result. The original problem can be stated as follows:

Suppose you have 100 gm of Martian potatoes, which contain 99 per cent water. Then you let them dehydrate till the water content is 98 per cent. How much do the potatoes weigh now? The answer is 50 gm. Isn't it surprising how such a small percentage reduction in the water content would imply such a large drop in the weight of the potatoes? Let us see how this happens. Let us start with 100 gm of potatoes. 99 gm of the potatoes is water. 1 gm is not-water. Let us dehydrate the potatoes till they weigh 50 gm. The amount of not-water remains the same. The amount of water is now 49 gm. 49/50 is 98 per cent.

The elevator paradox is a paradox which was first observed by Marvin Stern and George Gamow. Both of them were physicists who had their offices on different floors of a multi-storey building. Gamow, whose office was close to the ground floor, noticed that the first elevator to stop on his floor was the one going down most of the time. Stern, whose office was close to the top floor, noticed that the first elevator to stop at his floor was going up most of the time. It seemed as though elevator cars were being manufactured in the middle of the building and sent upwards to the roof and downwards to the basement to be demolished!

Obviously this was not the case. Then what could be the explanation for this strange observation? A simple explanation is that if we have one elevator, it spends most of its time in the larger section of the building, and is thus more likely to come from that direction. However, when there are more elevators the analysis becomes more complicated.

The 'interesting number paradox' is quite entertaining. If a number has some unusual property, it is said to be 'interesting'. For example, 2 is interesting because it is the only even prime number. 3 is the first odd prime and is therefore interesting. The sum of the proper divisors of 6 is equal to 6. That makes 6 an interesting number. Can we have numbers that are not interesting? The paradox states that all natural numbers are interesting. We can prove this statement by contradiction. Suppose there is a set of uninteresting numbers. The smallest uninteresting number would be of interest because it would be special in some way. That's why it is a paradox.

Have you heard about the missing square puzzle? It is actually an optical illusion and is used in Mathematics classes to help students reason about geometrical figures rather than draw conclusions based on outward appearances. It depicts two triangles made up of similar-looking pieces. One of the figures has a missing square.

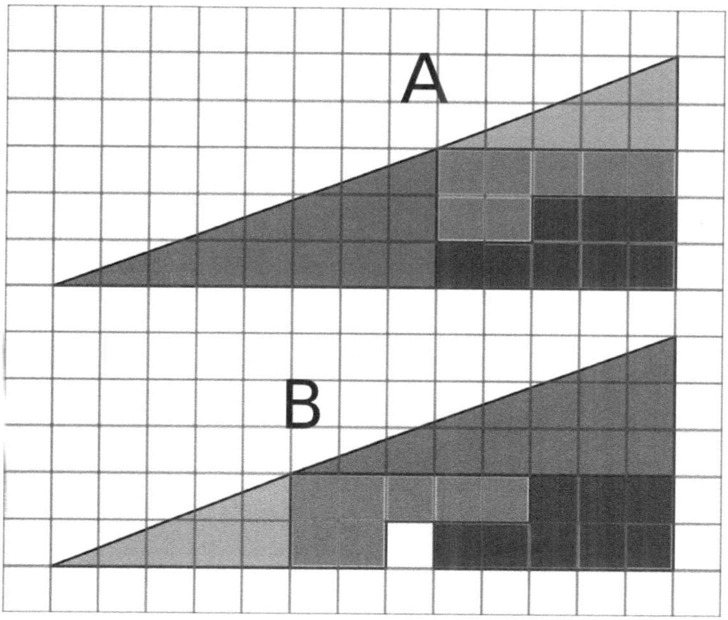

Figure 10.1: The missing square puzzle

The key to the puzzle is that neither of the two 'triangles' is truly a triangle, though they appear so to the human eye. The hypotenuse is slightly bent.

The human mind is fascinated by paradoxes. They are baffling and irresistible at the same time. There are many paradoxes of a philosophical nature—if God is all-powerful, why can't he create a stone, which he Himself can't lift? Let's go back to the age-old question—what came first—the chicken or the egg? There are many more such paradoxes. Can you create your own paradox?

11

FIBONACCI NUMBERS: NATURE'S SERIES

Look at the following series of numbers:

0, 3, 8, 15, 24, …

What comes next?

If you haven't guessed the answer, it is 35. Each number in the series follows a certain rule. If T_n is the nth term in the series,

$T_n = n^2 - 1$

Mathematics is filled with such series. At the basic level, these help children understand and apply the rules of arithmetic. At the advanced level, mathematicians, physicists and even engineers use different series to solve complicated problems.

Nature has its own series. The procreation of rabbits, the pattern formed by leaves on a stem, the arrangement of a pine cone and the family tree of honeybees can be described by numbers. These numbers form the Fibonacci sequence. It is an infinite series that starts with 0 or 1. The sequence is generated by the rule:

$$T_n = T_{n-1} + T_{n-2}$$

Thus, each term of the sequence is the sum of the previous two terms. The first few terms of the series are:

0, 1, 1, 2, 3, 5, 8, 13, … or

1, 1, 2, 3, 5, 8, 13, 21, …

Let us see how we can relate these numbers to the reproduction of rabbits. We have to start with a few assumptions. The first assumption is that rabbits start reproducing when they are two months old. Then they reproduce every month. Every pair of rabbits gives birth to a new pair. And they never die. Suppose we start with one pair of rabbits. After a month, there will be just one pair, but they would be mature. After two months, the rabbits would have reproduced another pair. We now have two pairs of rabbits. The next month, the initial pair would have reproduced again but not the second pair (because it would not be mature enough). So we'll have three pairs. Out of these, two are mature and one is immature. So the next month, there'll be five (three plus two) pairs. And so on. We can create a table using this method.

Month	*Immature*	*Mature*	*Total*
1	0	1	1
2	1	1	2
3	1	2	3
4	2	3	5
5	3	5	8

And the column indicating the total number of rabbits gives

us the Fibonacci series.

The surprising thing is that when we take any two successive Fibonacci numbers, their ratio is very close to the golden ratio. We came across this ratio in the chapter on pyramids. In Mathematics, two quantities are in the golden ratio if their ratio is the same as the ratio of their sum to the larger of the two quantities. When we express this algebraically, we say that:

$$(a + b)/a = a/b = \varphi$$

The Greek letter φ represents the Golden Ratio. Many artists and architects have used this ratio in their work, believing that this ratio has an aesthetic value. Its value is roughly 1.618.

What is even more surprising is that we can calculate any Fibonacci number using the Golden Ratio.

Another wonderful thing about the Fibonacci sequence is that if we take squares of widths given by the sequence and keep joining the opposite corners, we get a wonderful spiral.

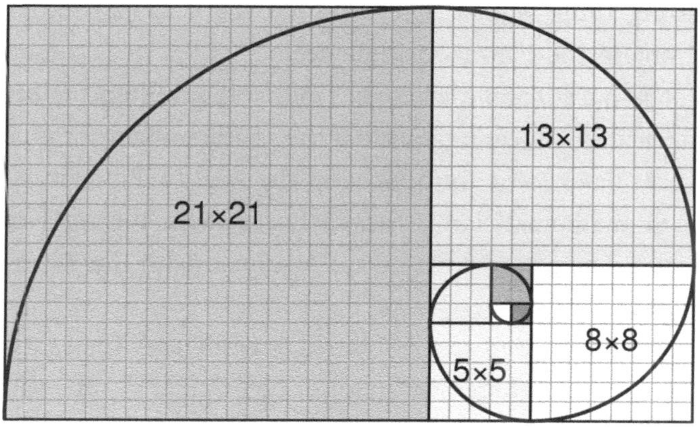

Figure 11.1: Fibonacci spiral

Have you heard of Pascal's triangle? It is a triangular array of numbers given by binomial coefficients. You would have read about algebraic expressions and coefficients. We come across binomial coefficients when we write down an expression for $(x + y)^n$. Here 'n' is a natural number.

When $n = 0$, $(x + y)^0 = 1$

When $n = 1$, $(x + y)^1 = x + y$

When $n = 2$, $(x + y)^2 = x^2 + 2xy + y^2$

When $n = 3$, $(x + y)^3 = x^3 + 3x^2y + 3xy^2 + y^3$

The coefficients of the various algebraic terms are the binomial coefficients. They form a triangle:

1

1 1

1 2 1

1 3 3 1

And so on. This triangle is known as Pascal's triangle. The amazing thing is that when we add the diagonal terms we will obtain the Fibonacci sequence.

It is questionable whether Fibonacci invented the series. Historical evidence suggests that the series probably originated in India, later to be introduced to the west by Fibonacci. The French mathematician Edouard Lucas, who invented other associated series, was the person who named it the Fibonacci sequence.

Where can we use the Fibonacci sequence? The sequence appears everywhere in nature, particularly in pattern formation—the number of petals on a plant, segments on a pineapple or a pine cone, the structure of nautilus shells can all be described by the Fibonacci sequence.

The series has a surprising application—it can be used while converting miles to kilometres. If you had to convert 5 miles to kilometres, you would obtain something very close to the number 8—which is the next number in the series. Sometimes while running a computer programme, we use the Fibonacci search, which is a method that uses the Fibonacci numbers.

If you are trying to find how many ways there are to tile a 2 × n grid with 1 × 2 dominoes, the answer is the nth term of the Fibonacci series.

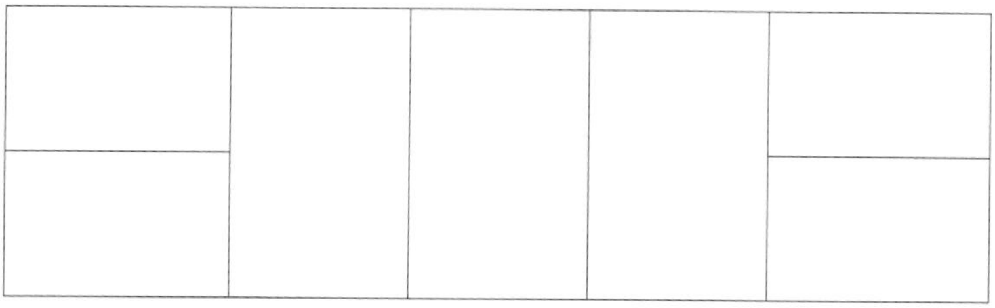

Figure 11.2: Example of tiling a 2×7 rectangle

If there are n stairs in a staircase and you can take them one or two at a time, the number of ways in which you can climb to the top is given again by the nth term of the Fibonacci sequence.

The series has several financial as well as mathematical applications. The Fibonacci sequence can be used by traders in a beneficial way. Isn't it amazing that a series, which gives the population of rabbits, can be used by traders? This is why Mathematics is so beautiful, and numbers are so full of mystery.

12

'THIS CAN'T BE HAPPENING!' THE MATHEMATICS OF IMPOSSIBLE OBJECTS

Take a good look at the figure below. What do you see? A pair of hands—each drawing the other! Now, is that something that could be possible? It isn't in 3D. But when we draw certain figures on a two-dimensional sheet of paper, we are able to depict impossible objects!

The artist Maurits Cornelis Escher who lived in the twentieth century is well known for his mathematically-inspired creations. Escher did not have any mathematical training. He just had a visual and intuitive idea of Mathematics. However, his creations display a strong mathematical influence. Some of the worlds that he drew were based on impossible objects such as the Penrose triangle and the Penrose stairs.

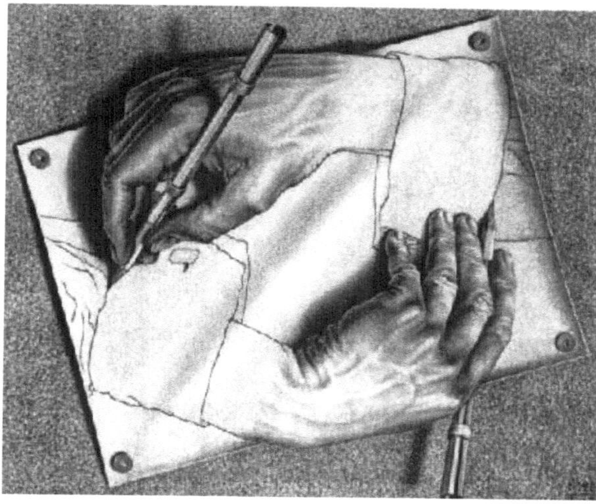

Figure 12.1: Escher's illustration

So, what is an impossible object? An impossible object (which is also referred to as an impossible figure or an undecided figure) is actually an optical illusion. It is a two-dimensional figure, which our visual system tries to interpret as a three-dimensional object and then concludes that the object is impossible. An impossible object comes as a bit of a shock, as we inherently desire to interpret 2D drawings as solid three-dimensional objects. Impossible objects are of interest to psychologists, mathematicians and artists and cannot be placed in just one of these categories.

The figure below is the drawing of a Necker cube. When we look at the drawing, our brain jumps to the conclusion that it is a cube, rather than two squares connected by diagonal lines. The left square seems to be facing us because we generally view objects from above. Focussing at a different point shows us that the right square could also be the one facing us.

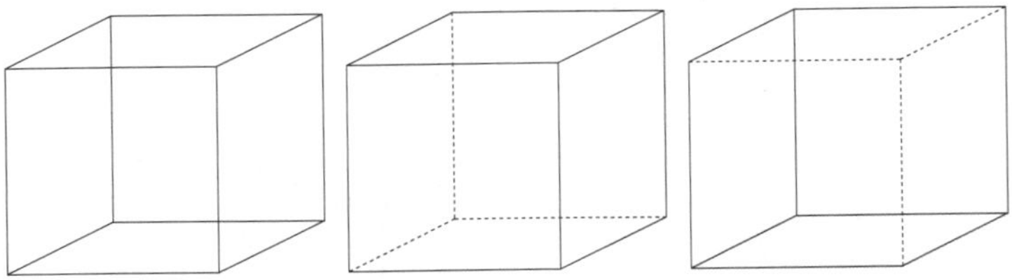

Figure 12.2: A Necker cube

This kind of uncertainty is known as ambiguity. M.C. Escher used this ambiguity to create the impossible cube, which seems to defy all laws of geometry!

Figure 12.3: The impossible cube

The Penrose stairs, also referred to as the Impossible Staircase, is an impossible object created by Lionel Penrose and his son Roger Penrose. It is a two-dimensional picture of a staircase that forms a loop. It gives the impression of a staircase that one can climb endlessly, but never reach any higher!

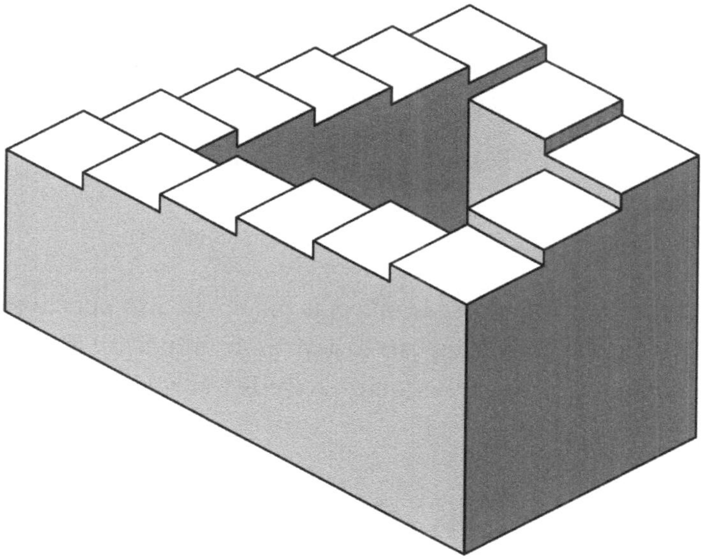

Figure 12.4: The Penrose stairs

The Penrose triangle is another example of an impossible object. It was created by the Swedish artist Oscar Reutersvärd in 1934. The psychologist Lionel Penrose and his son Roger Penrose (who was a mathematician) came up with the idea and popularized it in the 1950s, describing it as 'impossibility in its purest form'.

Figure 12.5: The Penrose triangle

A well-known impossible object is the blivet, also known as a devil's tuning fork. It appears to have three cylindrical prongs at one end which are mysteriously converted into two rectangular prongs at the other end.

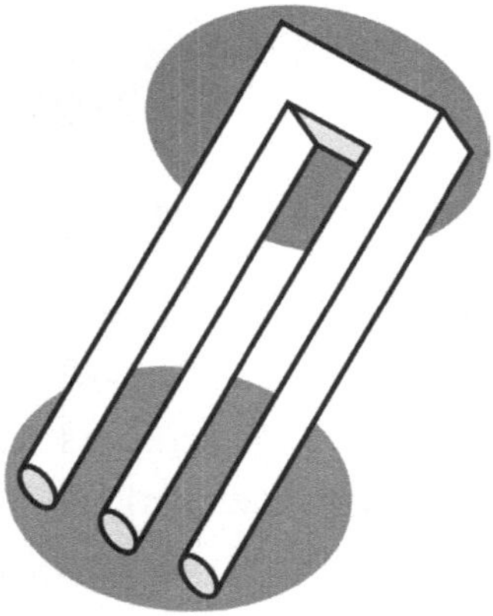

Figure 12.6: Devil's tuning fork

Impossible objects have been employed in science fiction, horror movies and books for special effects. For example, in the *Star Trek: The Next Generation* episode 'I, Borg', the *Enterprise* crew plotted a devious plan to eliminate the entire race of The Borg. Borgs were vicious aliens with interconnected minds. The plan was to show one of the Borgs a picture of a complex impossible object. This image would be transmitted to the Borg hive. The attempt to understand the image would overload their brains and subsequently destroy them. However, the plan was never carried out because it would amount to genocide or the deliberate killing of a large number of people of a certain race or community.

The TV series *The Simpsons* has made use of impossible objects for greater visual impact. In one of the episodes, the family runs through an Escher-like living room. In another episode, Lisa Simpson picks up a blivet from a garage sale.

Call of Cthulhu is a short story written by the American writer H.P. Lovecraft, where we have the city of R'lyeh. The city's architecture is based on non-Euclidean geometry (geometry which arises from a different set of assumptions) and looks quite impossible. Just gazing upon the city's architecture could cause one to be driven insane!

The TARDIS, the spacecraft and time machine in British TV serial *Doctor Who*, is also an impossible object. It is larger on the inside than on the outside! It is also based on non-Euclidean geometry. Non-Euclidean geometry seems to defy common sense. In Euclidean geometry or the normal kind of geometry, if we have a point A and a straight line, we can draw only one line that passes through A and is parallel to the given line. In non-Euclidean geometry it is possible to draw an infinite number of lines through A parallel to the given line.

In non-Euclidean geometry, the space itself is curved and we have elliptical and hyperbolic geometry.

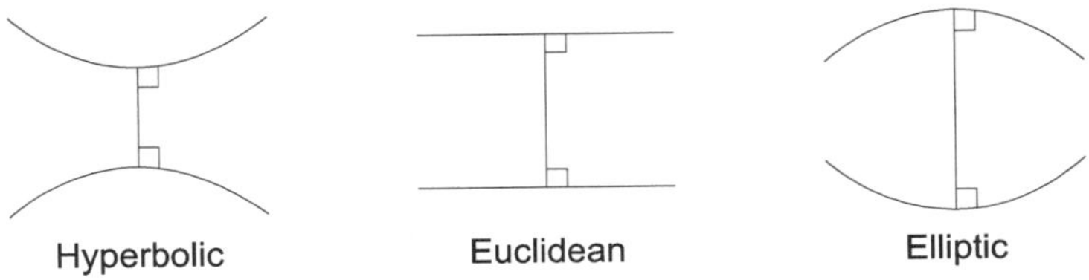

Figure 12.7: Euclidean versus non-Euclidean geometry

In order to explain impossible objects, we need both Mathematics and human psychology. Our world has three dimensions. When we view a three-dimensional object, a two-dimensional image is formed on our retinas. Our brains are able to relate this to a three-dimensional object. Now, certain two-dimensional figures exist, which cannot be related to objects in the three-dimensional world. These are the impossible objects. The paradox of the impossible figure is caused because the figure has two or more areas that are of interest to the human eye. Focussing on one area makes us believe that the reconstructed object is spatially aligned in a particular manner. Whereas focussing on another region, the eye concludes that the 'object' has another orientation. So when we look at the figure as a whole, it does not make any sense. As long as we look at the figure, there is disagreement in the two points of view, and we feel mind-boggled.

Artists love impossibilities. There was a movement called

surrealism that took place in the 1920s. Surrealists believe that the way to find truth in this world is through the subconscious mind and dreams rather than logic. Surrealism is all about experimenting with one's imagination. The movement included many artists, poets and writers. Surrealists drew their inspiration from Sigmund Freud's theories. Freud was a man who studied the mind, memories and human instincts. Surrealists loved to put together objects that you would not normally see together, and thus created art forms that bordered on the impossible.

'Persistence of Memory' created by the iconic surrealist painter Salvador Dali is probably the most famous piece of surrealist art. The painting depicts melting watches and gives one the sense that one is dreaming and that time is irrelevant.

Now that you have been introduced to the impossible, can you create your own impossible object?

13

WHY YOU WON'T BE ABLE TO READ THIS MESSAGE

You must have used codes to pass secret messages to your friends. Wasn't it fun? You would have probably taken a proper sentence, jumbled the letters of each word in some way and then let your friend unscramble it. What you did is known as 'encryption'. Did you know that encryption is an ancient science and has existed since people learned how to write?

The history of codes is fascinating. Codes were developed out of the need to pass on secret messages to one's friends without the risk of one's enemies being able to figure them out. Kings sent coded messages to their allies. George Washington, one of the founding fathers of the USA, sent encrypted messages to fellow soldiers. The Morse Code, which was developed when telegraph was invented in the nineteenth century, is an encryption system, which uses dots and dashes to represent each numeral and letter of the alphabet.

Throughout history we have had code makers and code breakers. Code makers would select a suitable method for encrypting a message. Code breakers would spend time trying to make sense of the scrambled message. Whenever code breakers cracked a particular method of coding, code makers would have to think extra hard to devise a new method of coding.

Today, in our computerized world, encryption has become much more complex and widespread. We use the term 'cryptography' instead of 'encryption'. Cryptography combines the disciplines of Mathematics, Computer Science and Electrical Engineering. When we use ATM cards, computer passwords or carry out electronic commerce, we rely on cryptography.

Let us study some of the easy ways of coding before we discuss what cryptography involves today. The two easiest methods of cryptography are transposition and substitution. In transposition, one takes a word, jumbles the letters and forms an anagram. For a small word, the method is not very effective because there are only a limited number of ways to jumble it. The word 'HAT' can be scrambled just to form 'HTA', 'ATH', 'AHT', 'THA' and 'TAH'.

Transposition works better for longer words. However, if the message is too long, it is almost impossible to decipher unless one knows the exact scrambling process. Transposition thus offers us a high level of security because the enemy interceptor would find it difficult to crack the code. But then, so would the recipient of the message. For transposition to be effective, the letters of the original message must be rearranged according to a rule that has been agreed upon by the sender and the receiver of the message.

Children sometimes use the 'rail fence' transposition technique to send messages to their friends. Here, we use two lines to scramble the message. Alternate letters are written in the upper and lower lines. For example, the following message:

THE ENEMY IS AT THE DOOR. BE PREPARED.

This is written as:

T E N M I A T E O R E R P R D

H E E Y S T H D O B P E A E

Then the second line is tagged onto the end of the first line to create the coded message. The encrypted message would be

TENMIATEORERPRDHEEYSTHDOBPEAE which sounds like gibberish. The process would have to be reversed to retrieve the original message.

The scytale, which is a military encryption device, also uses transposition. It is a wooden staff around which we wind a strip of leather or a parchment. The message is written along the length of the scytale. When we unwind the strip of leather, we get the encrypted message. In order to decrypt the message, we would have to wind it up again on a scytale having the same diameter.

Figure 13.1: The scytale, which is a military encryption device, also uses transposition.

Sometimes the messenger would wear the strip of leather as a belt, with the message written on the inside.

Instead of transposition, one can use substitution to send coded messages. Here, we substitute one letter of the alphabet with another. One recommended method of substitution is to randomly pair up alphabets and then substitute each letter in the message with its partner. If we take the English alphabet, the letters could be paired up as follows:

A C D G I L O P R T W Y B
Z E J K M S U B F N V H X

A is written as Z. Conversely, Z is written as A. C is written as E, E is written as C, D as J, J as D and so on. 'Let us attack at dawn' would be written as 'Scn ol znnzeg zn jzvt'.

A Caesar cipher, which is based on substitution, is a simple and one of the most commonly used methods of encryption. Here each letter of the alphabet is replaced by another letter, which is a fixed number of positions away from the initial letter. For example, when we have a left shift of 4, E is replaced by A, F would be written as B and so on. The method is named after Julius Caesar, the great Roman dictator, who used it to write private letters using similar techniques.

The methods we have discussed so far are simple and easy to decode. In the mid-nineteenth century, coding became much more complex. Information was classified according to the level of danger associated with it. This was known as the degree of sensitivity of the information. With the Official Secrets Act in 1889, it became illegal to reveal secret information. When the First World War started, secret information with varying degrees of sensitivity was being exchanged between various fronts. Coding became more sophisticated. People built machines to scramble and unscramble messages.

The earliest coding machine was the cipher disc, which had been invented in the fifteenth century by the Italian architect Leon Alberti. The cipher disc was made up of two discs, one slightly larger than the other. All the letters of the alphabet were inscribed at the edge of both the discs. The smaller disc was placed on top of the larger one and the two discs were fixed together with the help of a needle which acted as an axis. The letters on the outer disc represented the original text and the inner disc represented the cipher text. To create a coded message, each letter of the original message would be looked up on the outer disc and the corresponding letter on the inner disc would be substituted to form the coded message. To send a message with a Caesar shift of four, one would simply rotate the discs so that the A of the outer disc coincided with the E of the inner disc and the substitution process could be carried out.

Figure. 13.2: The cipher disc

At the end of the First World War, the German inventor Arthur Scherbius developed a cryptographic machine named Enigma, which was an electrical equivalent of the cipher disc. It turned out to be the most formidable system of encoding in history. This cryptographic machine had three basic elements connected by wires—a keyboard for inputting the original text, a scrambling unit which replaced each letter in the original text with a ciphertext letter and a display board, which consisted of lamps and displayed the encoded message. To decode the message the recipient had to know the exact settings of the machine.

Convinced that the Enigma code was unbreakable, the Germans used the machine for sending all sorts of classified information. The Polish were the first to crack the Enigma. In 1939, when Germany was about to attack Poland, the Poles shared their information with the British. The British then set up the Government Code and Cipher School at Bletchley Park in Buckinghamshire, where mathematicians and intelligence experts, with the help of early computers, were able to crack the Enigma code. This was one of the major factors that helped the Allied Powers win the Second World War.

An extremely innovative method of coding used during the Second World War was the use of Navajo Code talkers. The Navajos are a Native American tribe and their system of language bore no resemblance to any of the modern languages that are in use today and is quite incomprehensible to non-Navajos. During the world wars, Philip Johnston, a middle-aged civil engineer who lived in Los Angeles, came up with the idea that a secret code using the language of the Navajos could be used to transmit secret tactical messages. The US Marine Corps employed around two hundred Navajos for this purpose. They were known as Code Talkers. It was their task to translate English messages into codes, which were similar to their native language, and there would be Navajos at the receiving end who would decipher the message. This unique coding system proved to be very effective.

At the end of the twentieth century and in the beginning of the twenty-first century, there were giant leaps in the fields of telecommunication, computing hardware, software and data encryption. As computers became smaller, more powerful and less expensive, they became accessible to the home user and the small-business class. All the computers got quickly connected to each other through the Internet.

Since there was so much business being conducted through the Internet, such a lot of data being processed by computers and also random acts of terrorism, there was a real need for better methods for protecting the computers and all the information that they stored, processed and transmitted. Many academic disciplines such as Computer Security and Information Assurance were born out of this necessity.

Information Assurance or IA basically deals with all the threats present in the IT world such as viruses, worms, phishing attacks, identity theft, etc., and enables one to guard one's data from these threats.

Earlier encryption was for the benefit of the armies. Today, in our computerized world, we all rely on cryptography. Modern cryptography is advanced and complicated, but is based on the same principles as encryption, which is all about jumbling up letters.

14

NUMBERS COULD BE ENTERTAINING!

So far we have tried to fathom how big infinity is, delved into the mysteries of zero, discussed the significance of pi, the existence of imaginary numbers, learned how numbers can create chaos, found that we can use numbers to create impossible objects and so on. But the wonderful thing about numbers is that they can be interesting and can entertain us as well.

For example:

$111111111 \times 111111111 = 12345678987654321$

Isn't that cool?

The number 142857 is also special.

$142857 \times 1 = 142857$

$142857 \times 2 = 285714$

$142857 \times 3 = 428571$

$142857 \times 4 = 571428$

$142857 \times 5 = 714285$

$142857 \times 6 = 857142$

Can you see how the sequence repeats itself?

Also:

142 + 857 = 999

14 + 28 + 57 = 99

142857 × 142857 = 20408122449

On adding,

20408 + 122449 = 142857!

There are many big numbers, which are interesting in some way.

86400 is equal to the number of seconds in a day.

31556926 is equal to the number of seconds in a year.

7,000,000,000 is the estimated human population on Earth.

One googol (a misspelling of which is how Google got its name!) is the number 1 followed by 100 zeroes.

One centillion is 1 followed by 303 zeroes.

One googolplex is 1 followed by googol zeroes. The only problem is that there is not enough room in the universe to write this number.

Did you know that numbers could be amicable? Amicable numbers are a pair of numbers that love each other! And what do we mean by "love each other"?

Consider the numbers 220 and 284.

The proper divisors of 220 are:

1, 2, 4, 5, 10, 11, 20, 22, 44, 55, 110.

The proper divisors of a number include all the factors of a number excluding the number itself.

On adding all these numbers, we get 284.

The proper divisors of 284 are:

1, 2, 4, 71, 142

Adding them, we obtain 220. That's how 220 and 184 are a pair of amicable numbers. Other amicable pairs include (1184, 1210), (2620, 2924) and (5020, 5564). Amicable numbers were discovered by the followers of Pythagoras and have been researched through the centuries by mathematicians.

Another interesting concept is an 'emirp'. We obtain this word by spelling 'prime' backwards. True to its name, the word refers to a prime number that becomes a new prime number when you reverse its digits. One-digit prime numbers or palindromic primes (such as 151 or 787) do not fall in this category. The first few emirps are 13, 17, 31, 37, 71, 73, 79, 97, 107, 113 and 149. Reverse them and see what you get.

Some numbers have been declared to be happy. In order to find out if a given number is 'happy' or not, we need to perform certain operations on them. Consider the number 32. We square each of the digits and add them:

$3^2 + 2^2 = 9 + 4 = 13$

Repeating the same procedure:

$1^2 + 3^2 = 10$

$1^2 + 0^2 = 1$

Whenever we perform these operations and the final result is 1, the original number is said to be happy. 32 is thus a happy number. When the final answer isn't 1, the number is declared to be unhappy. All unhappy numbers follow sequences that reach the cycle:

4, 16, 37, 58, 89, 145, 42, 20, 4

There are lots of happy numbers. There are 11 happy numbers between 1 and 50. The largest happy number that has no repetitive digits is 986,543,210. What a happy thought!

We also have perfect numbers. In number theory, a perfect number is a positive integer that is equal to the sum of its proper divisors. The first perfect number is 6. The proper divisors of 6 (1, 2 and 3) add up to give us 6. The next perfect number is 28. $1 + 2 + 4 + 7 + 14 = 28$.

496 and 8128 are the next two perfect numbers. The first four perfect numbers were the only ones known to ancient Greek mathematicians. Philo of Alexandria (an Alexandrian philosopher who combined Judaism and Greek wisdom in his philosophy) wrote the book *On the Creation* in the first century AD where he claimed that the world was created in 6 days and the moon orbits around the earth in 28 days because both 6 and 28 are perfect numbers! As of 2016, there are 49 known perfect numbers.

Have you heard of Narcissus? He was a hunter from Greek mythology. He was extremely beautiful and also very proud. When he saw his own image in a pool, he was so mesmerised by it that he fell in love with it, as he didn't realize that it was just a reflection. He eventually lost the will to live or do anything but stare at the image. Transfixed, he stared at his own reflection till he died. Well, we have narcissistic numbers as well. They are also known as Armstrong numbers or plus perfect numbers. A narcissistic number equals to the sum of each of its digits when these digits are raised to the power of the number of digits in the number.

Let me explain:

153 is a narcissistic cube. Why? The number has three digits. So, we raise each digit to the power 3 and add them.

$1^3 + 5^3 + 3^3 = 153$.

370, 371 and 407 are the other three narcissistic cubes.

There are no one-digit narcissistic numbers or 12-digit or 13-digit narcissistic numbers. Although people have spent time and effort hunting for these numbers, English mathematician G.H. Hardy realized that the whole exercise was fun but a silly pursuit. He stated in his book *The Mathematician's Apology* that 'these are odd facts, very suitable for puzzle columns and likely to amuse amateurs, but there is nothing in them which appeals to the mathematician.'

We also have 'weird' numbers. The sum of the proper divisors of a weird number is greater than the number, but no subset of the divisors add up to give the number itself. For example, the smallest weird number is 70. Its proper divisors are 1, 2, 5, 7, 10, 14 and 35. Their sum is 74 (greater than 70), but no subset of these sums upto 70.

An 'untouchable' number is a positive integer that cannot be expressed as the sum of all the proper divisors of any positive number, including the positive number itself. For example, the number 4 is not untouchable as it is equal to the sum of the proper divisors of 9 (1 + 3 = 4). The number 5 is untouchable because it cannot be written as the sum of the proper divisors of any positive integer.

You have read about 'interesting numbers' and seen that the first dull number would be interesting. So the category of dull numbers goes out of the window. For people who love numbers, all numbers are interesting!

15

IS ALL MUSIC ACTUALLY MATH?

The *Concise Oxford English Dictionary* defines music as 'the art of combining vocal or instrumental sounds to produce beauty of form, harmony and expression of emotion'. Mathematics, on the other hand, is the study of numbers, shapes and space using reason, and a special system of symbols and rules to organize them. It is amazing how two such diverse fields could have so much in common. Mathematics and music are intertwined and we can conclude that, in the larger picture, music is just an applied form of Maths.

Consider Beethoven, the great composer, who was probably the most famous and highly regarded amongst all composers. His hearing began to deteriorate at an early age. He spent the last decade of his life practically deaf. He stopped performing and conducting music in public, but he continued to compose music and some of his best pieces of music were composed during this period of time. How did he manage to do that? The answer lies in the correlation between music and Maths.

A note can be represented by a certain mathematical pattern. Some notes combine to form beautiful patterns that repeat themselves. The music created by such patterns sounds very pleasant. To create beautiful music, one would need repetitive patterns with elements of surprise. Beethoven was able to see and feel musical patterns and was probably able to compose music using both his instinct for music and the use of Mathematics for precision.

People have also tried to find connections between the music created by Mozart and Mathematics. Wolfgang Amadeus Mozart was born in Salzburg, Austria. He lived in the late eighteenth century and died when he was just 35. He was a child prodigy, and in his short life, he managed to write over six hundred brilliant pieces of music that are considered to be some of the best compositions from the classical era of western music. Scholars have been studying Mozart's music for a long time and are trying to figure out *how* he composed such brilliant music. Was it his musical genius at work or did he rely on mathematical equations?

Astrophysicist and author Mario Livio, who studied the relationship between music and Mathematics, is convinced that Mathematics played a role in Mozart's compositions. There is symmetry in Mozart's compositions. Harmony and symmetry are present in Mathematics too. Some scholars believe that Mozart used the golden ratio and Euclidean Mathematics in his compositions. It is well known that Mozart carried music in his head and that he often jotted down mathematical equations in the margins next to his compositions. These do not directly prove the conjecture, but there is enough evidence to show that Mozart was drawn to Mathematics and numbers. We can only speculate!

We have been talking about patterns in Mathematics and music. Let us see how they arise. You know that sound is a sort of wave where air molecules vibrate at a certain pitch or frequency and cause their neighbouring molecules to vibrate too. Finally, vibrating air molecules impinge on our eardrums. Our hearing apparatus picks up these vibrations and sends a message to our brain, which interprets it as sound. Now, a pure note is a vibration of a single frequency. It can be mathematically represented by a sine wave, which looks like this:

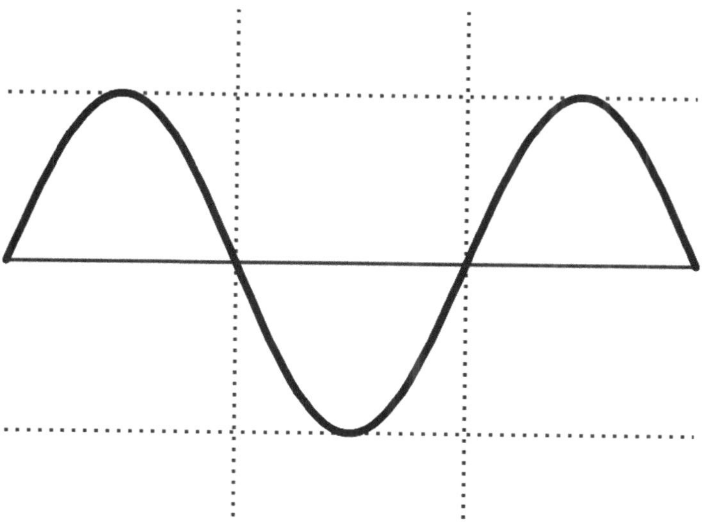

Figure 15.1: Vibrating air molecules

This diagram tells us how much the air molecules have moved and in which direction in relation to time. Sound waves have a frequency range from 2,000 to 20,000 cycles per second. Music consists of notes played in succession. These produce more intricate patterns.

Then there is a lot of Mathematics involved in the design of string instruments. The frequency of a string instrument is given by a formula that links it to the length of the string, tension in the string and mass per unit length.

Shorter strings vibrate with a higher frequency. As a musician presses a string, he or she effectively shortens it. The greater the number of fingers used, the shorter the string becomes, producing even higher frequencies.

Tension in the string is related to how tightly the string is stretched. A tightly stretched string will produce a higher frequency and a loosely stretched string will produce a lower frequency of vibration. The mass per unit length of the string or 'm' indicates the density of the string. Strings having higher densities tend to produce sound with lower frequencies as opposed to strings having lower densities.

Wind instruments like flutes can also be studied using Mathematics. Mathematics and rhythms have a lot in common. The Euclidean algorithm, which is an efficient method of finding the highest common divisor of two numbers, can be used to generate all the traditional musical rhythms.

There is more Mathematics in music. The pitch represents a certain frequency. If we double the frequency we will hit a note that is one octave higher.

Many composers have used phi or the golden ratio in the

timing of their compositions. The climax of many compositions occurs at roughly the phi point of the song, as opposed to the middle or the end of the song.

People have used fractals to create music, which is not surprising, since repetitive themes and self-similarity are present in music.

A composer carries out the task of fitting chords and melodies together to form a harmonious whole and creates a beautiful piece of music. This sounds as though the basic elements of music have geometrical shapes that can somehow be fitted together. Recently, Dmitri Tymoczko, a Princeton University professor of music, has used advanced Euclidean geometry and has shown that we can actually visualize music being 'constructed' in an abstract space. His theory might provide a tool for composers who are trying to figure out the next chord in a particular composition.

Mathematics and music have more in common. Research has shown that listening to classical music can enhance one's mathematical ability. There is also something known as the Mozart effect, where people who have been listening to Mozart's compositions for an extended period of time can improve one's spatial-temporal abilities, which is needed in Mathematics. Spatial-temporal ability has to do with fitting objects together in space. Solving jigsaw puzzles is an example of this ability.

Oh, and mathematicians and physicists love music. Einstein was a talented violinist and had a lifelong passion for music. Max Born, the German mathematician and physicist who was instrumental in the development of quantum mechanics, loved to play Bach.

16

ALL THAT THE PLANET NEEDS IS FOUR COLOURS

You have just read about the connection between music and Mathematics. Another surprising place where Mathematics creeps in is when we are trying to colour a map. Let's go back to the basics. Let's take a pattern—any pattern, and call it our map. How many colours would we need to use to colour it? Here, we would have to stick to one basic rule: two regions (the regions could be countries or states) that have a common boundary cannot be coloured the same. Two regions having a common corner can be coloured similarly, but not two regions having a common boundary.

Let's start with a straightforward pattern. Let's take a group of sixteen squares:

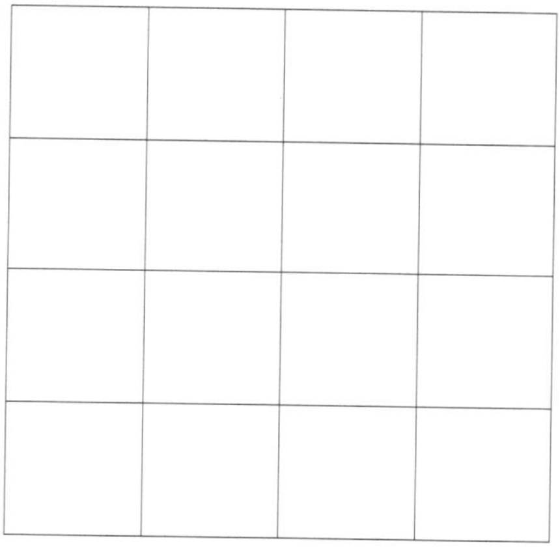

Figure 16.1: A group of sixteen squares

How many colours would we need to colour this pattern? We could use sixteen different colours, but two are sufficient!

Figure 16.2: A group of sixteen coloured squares

Let us now take a slightly more complicated figure. Let us

take a circle surrounded by another circle and divided into four similar regions:

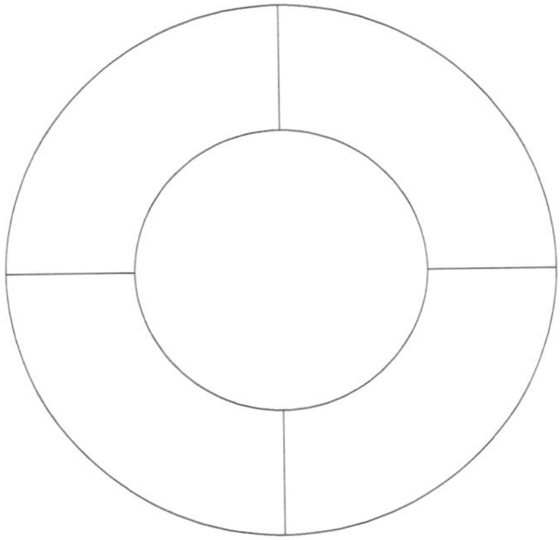

Figure 16.3: Concentric circles

How many colours would we need now? We could use five different colours, but three are sufficient.

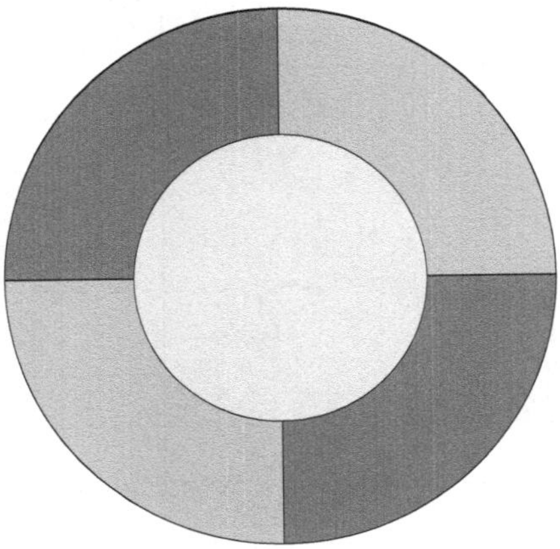

Figure 16.4: Coloured concentric circles

If we have an odd number of regions surrounding a circle, four different colours are needed.

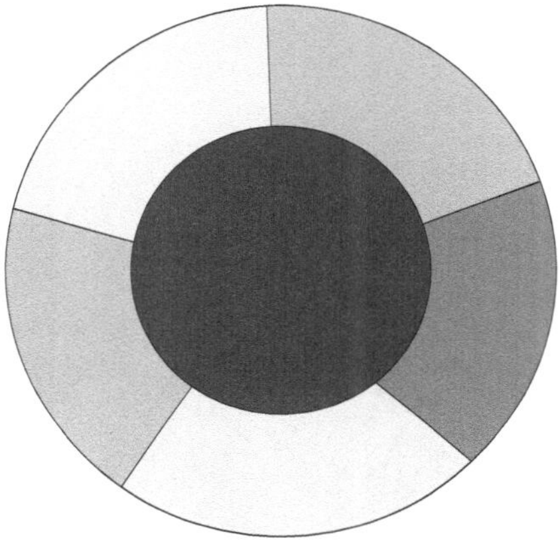

Figure 16.5: Concentric circles with four different colours

Now take a look at the map of Madhya Pradesh and its five neighbouring states:

Suppose the map was drawn on an elastic material such as a rubber. By stretching and reshaping the map, we could have transformed the map into Figure 16.5. Figure 16.5 and Figure 16.6 are said to be homeomorphic. So we are able to colour the map of Madhya Pradesh and its neighbours using four colours. We can colour the map of entire India using four colours.

It seems that any pattern or map can just be coloured with four colours. This may not work if a country has two or more separate areas. For example, Alaska is a part of the US, but there is Canada in between. Let's ignore such cases. In some cases, we would need less than four colours. In many cases, we can use as many colours as we want but four are enough. This theorem—known as the four colour theorem—is one of the most famous theorems of Mathematics.

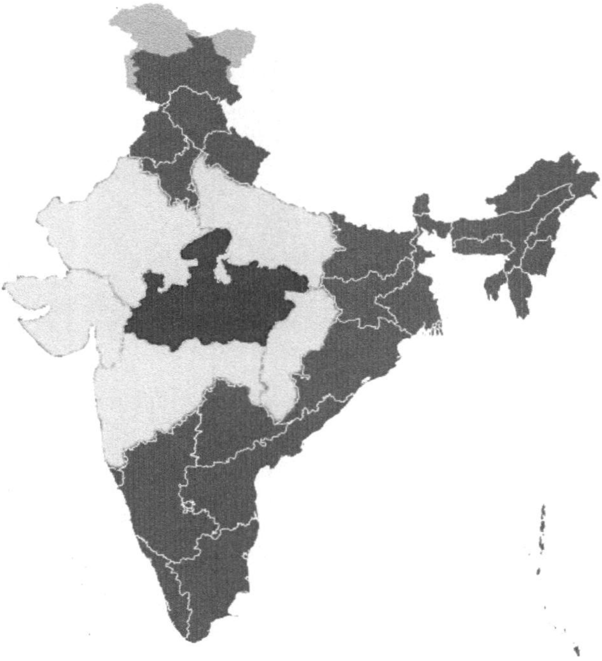

Figure 16.6: Madhya Pradesh and neighbouring states

The theorem is important because it was first stated in 1852, but no one was able to prove it until 1976. For over one hundred and twenty years some of the brightest mathematicians racked their brains over this seemingly simple theorem, but were unable to come up with a proof. People did offer solutions, but these solutions turned out to be erroneous. An entirely new branch of Mathematics—known as Graph Theory—was developed in order to solve the problem. It was only in 1976 that Kenneth Appel and Wolfgang Haken proved the theorem with the aid of a computer. It was the first theorem to be solved with the help of a computer. Some purists objected to the idea of a computer being used to prove a mathematical theorem. Would you agree with them?

Here is a map of the US with its various states. Can you colour it using four colours?

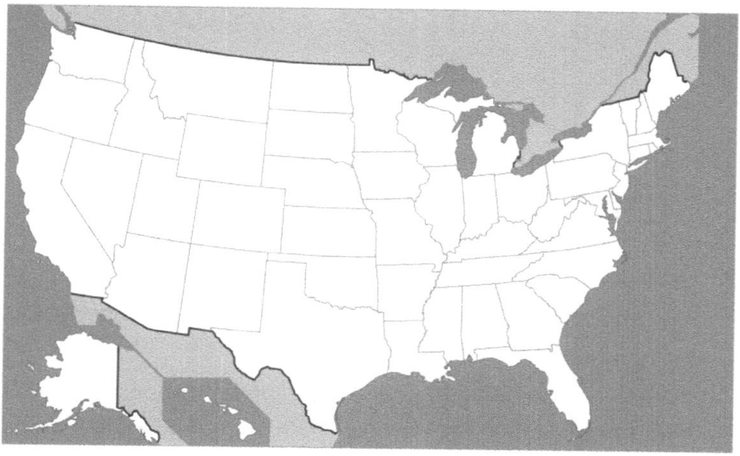

Figure 16.8: Map of the US

17

HOW THE WHOLE WORLD IS CONTAINED IN JUST TWO NUMBERS

Think of the early man. His brain was just developing, he had learned to speak and he felt the need to count objects, days, people and so on. What would he have done? He would have used his ten fingers for counting. That is why we have the decimal system, and 10 has been the basis for most of the counting systems throughout history.

When the early man needed to record numbers, he would have made scratches on the cave wall or notches on a stick. Ancient records show that numbers were broken up into groups of tens and ones. There was a symbol for 1 and a symbol for 10. There is evidence to show that ancient Egyptians (who lived around 3000 BC) represented 1 by a vertical line and 10 by the symbol ^. Thus, the number 23 was written as III^^.

People have used other numbers as bases for their number systems. The Babylonians used the number 60. They would have needed 59 different symbols to represent numbers up to

59. They solved this problem by expressing numbers below 60 in clusters of ten. However, the written symbols were not well suited for mathematical computation. But the Babylonians had studied astronomy in detail, and even today, we use 60 as a base in some measurements. For example, there are 60 seconds in a minute; the angles of a triangle add to 180 degrees, and there are 360 degrees in a circle (180 and 360 are multiples of 60).

The Babylonians also introduced the concept of place value—which means that a particular digit has a different value depending on its place in the sequence. For example, in the number 555, each 5 has a different meaning. The last 5 denotes 5, the second 50 and the first 5 denotes 500.

The zero, the decimal system and the way numbers are written today evolved in India around AD 800. These then reached Europe two hundred years later through their documentation in Arabic manuscripts and thus came to be known as Arabic numerals. The decimal system is now used all over the world.

Now imagine a number system that has only two digits—0 and 1. How would you count objects using this system? You would count 0, 1 and then 10. After counting 11, you would reach 100. Sounds a bit ridiculous, doesn't it? Well, such a number system exists and is known as the Binary (meaning 'two' or 'a pair'!) System, and it's used in computers! I will explain further, but let us first explore what the binary system is. Here are the binary equivalents for the first eight decimal numbers:

DECIMAL NUMBER	BINARY EQUIVALENT
0	0
1	1
2	10
3	11
4	100
5	101
6	110
7	111

The place values in binary numbers are different. In the decimal system, we have units, tens and hundreds. After the decimal point, we have 1/10ths, 1/100ths and so on. In the binary system, we have twos, fours, eights, sixteens. And after the decimal point we have halves, quarters, one-eighths, etc.

Consider the binary number 1100101.1001. Let us try to calculate its decimal equivalent. The digits on the left of the decimal point indicate the integer part and the digits on the right indicate the fraction part of the number.

The binary number 1100101.1001 is thus equivalent to:

$1 \times 1 + 0 \times 2 + 1 \times 4 + 0 \times 8 + 0 \times 16 + 1 \times 32 + 1 \times 64 + 1 \times 1/2 + 0 \times 1/4 + 0 \times 1/8 + 1 \times 1/16 = 1 + 4 + 32 + 64 + 0.5 + 0.0625 = 101.5625$.

How do we indicate that a particular number is binary? We add a subscript of 2 to the number. For example: $1010_2 = 10$. The modern binary number system was invented by the great polymath and philosopher Gottfried Leibniz, who lived in the seventeenth century. (A polymath is someone who is an expert on various, seemingly unconnected, fields.) Along with several other contributions to the field of Mathematics, Leibniz worked independently of Sir Isaac Newton and developed calculus.

Leibniz believed that logic or 'laws of thought' could be mathematically stated. He also had ideas of constructing a universal language that could express all conceptual ideas. Such a language would be extremely difficult to construct, but he hoped that it would be understood by all. It was a high-flown idea at that time, but this is the very idea that lies at the heart of modern computing.

Computers today use the binary system to represent information, where computer software acts as a translator between the binary information and the information we actually work with. This could be text, photos, sounds or videos. Binary language is also known as machine language because it is the language the computer can understand. Each 0 or 1 is called a bit or a binary digit. One bit doesn't convey much. So bits are grouped together in groups of eight, known as bytes. A byte can represent numbers from 0 to 255.

When data has to be transmitted over a telephone line or a radio link, high- and low-pitched tones are used to represent a series of 1s and 0s.

Audio CDs, CD-ROMs and DVDs store bits optically, which means that we have a spinning disc on which data has been recorded, in a pattern that can be read using a beam of laser light. The reflecting part of the disc might represent a 1 and the non-reflecting part could indicate a 0. No matter where information is stored in an electronic device, at the fundamental level, it consists of 0s and 1s. A computer stores numbers, text and all other information in the form of bits and bytes. A binary code is needed to make sense of the data.

Computers use logic gates to process information, perform mathematical operations and so on. A logic gate might sound terribly complex, but it is basically an electric circuit with two inputs and one output. It receives the two incoming currents, compares them and then sends an outgoing signal depending on the incoming signals and the kind of gate it is. Binary logic and binary numbers are used for these logic gates as well.

Can you imagine life without computers and other electronic devices? Everything that a computer processes and displays on the screen has a language of two numbers as its basis! Binary is the language these machines understand. For these devices, the whole world is contained in just two numbers.

18

WHY THE FINAL MYSTERY WILL ALWAYS BE BEYOND US: GÖDEL'S INCOMPLETENESS THEOREM

In the year 1931, a young mathematician named Kurt Gödel made a huge discovery. The idea that he put forward was as powerful as any of the theories developed by Albert Einstein. While Gödel's discovery basically pertained to the field of Mathematics, it can be applied to all branches of Science, logic and human knowledge. In fact, it changed the very way people look at the world.

Let me tell you a bit about Kurt Gödel. He was born on 28 April 1906 in Brno, which is now in the Czech Republic but at that time was a part of Austria-Hungary. His father was a well-to-do textile manufacturer. From all accounts, Gödel seemed to have had a happy childhood. Apparently, he was a very curious child and his inquisitive nature earned him the name 'Mr Why'. He was interested in Mathematics and

Philosophy from a very early age, and by the time he was seventeen, he had mastered university-level Mathematics. He joined the University of Vienna to study theoretical physics but shifted to mathematics and then to mathematical logic. Gödel joined the faculty of the university after receiving his doctorate. When he was twenty-five, he published the famous proof of the Incompleteness Theorem.

What exactly was Gödel's Incompleteness Theorem? To understand the theorem, you must appreciate the fact that mathematicians *love* proving things. They were agitated for centuries because they could not formally prove some of the things they intuitively knew were true. Geometry at the high school level consists of theorems and proofs, based on which one studies triangles and parallelograms and proves more things!

The five postulates (i.e., suggestions or assumptions) of Euclid form the basis of high school geometry. Everyone knows that these postulates are true, but nobody has been able to formally prove them. For example, the first postulate says that if we have two points, we can always draw a straight line joining them. But has anyone been able to prove it? The answer is *no*. So many mathematical minds have been completely frustrated for over two thousand years because they have been unable to prove all their hypotheses.

However, in the early twentieth century, mathematicians began to feel extremely optimistic. There were some brilliant mathematicians—Bertrand Russell, David Hilbert and Ludwig Wittgenstein, to name a few—who were convinced that they would soon be able to present a grand theory which would prove everything in Mathematics. And Mathematics would emerge as a magnificent, complete and bulletproof field of study.

In his paper, Gödel proved that we would *never* be able to formulate a single theory that would explain everything in Mathematics! Gödel's discovery dashed the hopes of all the eminent mathematicians of his times.

In fact, Gödel stated and proved two theorems about the incompleteness of any mathematical model. Only full-fledged mathematicians will be able to understand these theorems if they are stated in formal language. So I won't state them formally here. I'll just give a gist of what they actually mean.

Gödel's theorems tell us that if we draw a circle around a set of things, we cannot fully explain them without referring to something outside the circle. For example, you can draw a circle around all the theorems of high school geometry. But they are built on Euclid's five postulates, which are true, but cannot be proven within the circle.

You can draw a circle around all the different types of cars. But the existence of these cars relies on factories that manufacture them. A car cannot explain itself! Gödel proved that there will always be more things that are true than you can prove. Any logical or mathematical system will be based on certain assumptions and some of them will remain unprovable.

Gödel used a highly complex chain of reasoning to prove his point. Though Gödel's Incompleteness Theorem is mathematical, it applies to *everything* that depends on logic or reasoning. It is true in Science, Language and Philosophy. The theorem basically tells us that human knowledge will never be complete.

Gödel used the Liar's paradox to prove his point. The statement 'I am a liar' is self-contradictory. If it is true, I am not a liar. So it is false. If it is false, I am a liar. So it is true!

With the help of this example, Gödel proved that any statement requires an external observer. No statement alone can prove itself to be true. Thus, any mathematical system needs something outside itself to prove the truth of the statements in the system.

It is interesting to note that Gödel moved to the Institute for Advanced Study in Princeton University, New Jersey, in 1933. There he met Einstein for the first time. Gödel and Einstein discovered that they were intellectual equals and because of their shared cultural background as well, they got along well. They met and talked to each other almost daily from 1942 until Einstein's death in 1955.

Physics is an area where Gödel's theorem can be successfully applied. Philosophers and scientists, starting with Aristotle who lived in the fourth century BC, have been trying to understand the fundamental forces of nature in a qualitative manner, which means that people used adjectives to describe things rather than numbers. Newton gave precise and mathematical laws of motion and gravitation. People began to think that if they could find the position and velocity of each particle of the universe, they would be able to predict the future of each particle and thus of the universe.

Then, in the early twentieth century, it was discovered that radioactivity takes place in a completely random manner, which challenged the scientific community. But then, Quantum Theory was discovered. Quantum Theory deals with the motion of particles at the atomic and sub-atomic levels. The theory goes on to suggest that there is randomness and unpredictability at these levels.

Physicists have been trying to discover all the laws of nature operating at the quantum level. They hope to find a grand Theory of Everything one day. But many scientists are also of the opinion that a grand theory of everything will never be discovered. That is because we are trying to create a mathematical model of the universe with a finite number of principles.

Gödel's theorem proves that we cannot do that. However, we should not abandon our quest for knowledge. We may never find the whole truth, but we can come closer and closer to it.

19

INDIA'S CONTRIBUTION TO MATHEMATICS

We now know from Gödel's theorem that our knowledge of Mathematics will never be complete. But that should not prevent us from trying to understand more. Mathematics is all around us. It is an essential part of our lives and is built into everything that we do. Computers, electronics, architecture, art, money, engineering and even sports rely on Mathematics.

There are two kinds of Math—Applied and Pure. Applied Mathematics is used to study the physical, biological and the social world. Applied Mathematics makes it easy to solve problems in science. Scientists build a mathematical model for any phenomenon, solve the equations in the model and then try and make predictions based on the model. They also try to improve the performance of a system with the help of a mathematical model.

Pure Mathematics, on the other hand, deals with abstract concepts rather than real world phenomena. These abstract concepts have led to major discoveries for mankind. The Universal Turing machine, for example, was an abstract idea formulated by Alan Turing in 1937. It led to the development of the modern computer.

Every society, whether civilized or primitive, has used some form of Mathematics. The need for Mathematics arose from the needs of the society, and as societies grew in complexity, so did their mathematical needs. The early man used Mathematics to count objects and to calculate the position of the sun. One of the first counting systems was developed by the Sumerians. Then the subject of arithmetic with basic operations such as addition, subtraction, multiplication, division and square roots were developed. The Mayans in the US had an elaborate calendar system and knew about astronomy. They also knew about the number zero, but were not very clear on how they could utilize it.

It was in ancient India that zero was recognized as a number and the rules of its use were defined. As civilization progressed, the subjects of geometry and algebra were introduced. Geometry deals with various shapes, areas and volumes, and with it comes many practical applications. With the help of algebra, people were able to divide inheritances and use resources in a judicious and optimum way.

Ancient mathematicians also began to look at Number theory, which is all about the properties of positive numbers and the relations between them. Mathematics flourished under the Greeks. They studied and developed abstract concepts in Mathematics. The beautiful architecture of the Greeks and their complex system of government were based on their knowledge of Mathematics. Some of the earliest philosophers and mathematicians—Pythagoras, Zeno, Aristotle, Plato, Archimedes and Euclid, to name a few, came from this era. During this time, trigonometry was also developed.

Trigonometry deals with the measurement of angles and the computation of trigonometric ratios related to them. Trigonometry could be applied to astronomy and in the study of heights and distances.

After the fall of the Roman Empire, the Arabs added to the knowledge of Mathematics. Fibonacci was one of the first European mathematicians. The Renaissance in Europe led to advancement in many areas of Mathematics.

In the seventeenth century, Sir Isaac Newton and Gottfried Leibniz independently developed the basis of calculus. Calculus is an advanced form of Mathematics and is of great help to physicists, engineers and mathematicians.

So that was the story of the evolution of Mathematics till the seventeenth century. Mathematics has developed further since then and most of it is terribly complex. What I would like to discuss here in particular is India's contribution to Mathematics.

There were significant advances made in Mathematics in India from 1200 BC to the end of the eighteenth century. During the classical period of Indian Mathematics (from AD 400 to AD 1600), mathematicians such as Aryabhatta, Brahmagupta, Mahavira, Madhava of Sangamagrama and Nilkantha Somayaji made some very important contributions.

The decimal system was invented in India. Early Indians discovered the benefits of a decimal place value system and were using it before the third century. The Chinese had also thought of it, but the Indians refined and perfected the system. The nine numerals that we use worldwide today were derived from their Indian ancestors.

The concept of zero as a number in its own right was also introduced in India. The seventh century Indian mathematician Brahmagupta established the basic rules for dealing with 0 as well as negative numbers. He also pointed out the fact that quadratic equations (equations having squared terms in them, for example, $x^2 - 4 = 0$) could have two possible solutions and one of them could be negative. He also tried to write down these concepts, using initials of the names of colours as unknown variables in his equations. That is similar to what we do in algebra today. We use 'x', 'y' and 'z' to denote unknown variables.

During the Golden Age of India, when the Guptas were in power (from the fourth to the sixth century), Indian mathematicians made many advances in the field of trigonometry. Trigonometry links geometry and numbers and was first developed by the Greeks. Indian mathematicians used trigonometry to survey the land, navigate the seas and also chart the skies. Indian astronomers used trigonometry to calculate relative distances between celestial bodies. In trigonometry, one comes across trigonometric ratios. The Greeks had been able to evaluate the trigonometric ratios of certain angles. Indian scholars were able to evaluate the trigonometric ratios for all angles.

Bhaskara II, who lived during the twelfth century, was one of the greatest Indian mathematicians. He was the first mathematician to develop the concept of division by 0. He noticed that when we divide 1 by smaller and smaller fractions, we obtain larger and larger numbers. Thus, dividing a number by zero would yield infinity.

The Kerala School of Astronomy and Mathematics was founded in the late fourteenth century by Madhava of Sangamagrama, who was probably the greatest mathematician and astronomer of Medieval India. He developed infinite series for a large number of trigonometric functions and π. It is possible that the development of calculus in Europe was influenced by his work.

It is evident that negative numbers, arithmetic, algebra and trigonometry were all thought up by Indian mathematicians. Then, these ideas were transmitted to the Middle East, China and Europe.

Mathematical rules and problems were usually stated in the form of short verses known as 'sutras' to help people memorize them. During the eighth century BC, long before Pythagoras's time, a set of verses known as the 'Sulba sutras' gave the general statement of Pythagoras' theorem and also listed several simple Pythagorean triplets such as (3, 4, 5), (5, 12, 13), (8, 15, 17) and (7, 24, 25).

Excavations in Mohenjodaro and Harappa show us that ancient Indians made use of practical Mathematics. The proportion 4:2:1 was considered favourable for the stability of a brick structure. They had their own standard system of weights based on the ratios 1/20, 1/10, 1/5, ½, 1, 2, 5, 10, 50, 100, 200 and 500. They had also designed a ruler for the measurement of length.

Baudhayana, who lived in the eighth century BC, gave a formula for the square root of 2. Sanskrit grammarian Panini formulated Boolean Logic (the binary logic that is used in computers), and Pingala, a music theorist, quite by chance discovered Pascal's triangle and binomial coefficients.

Ancient Indians did not differentiate between Mathematics, poetry and religion. There were mantras, which invoked the powers of ten from hundred to trillion. A fourth century Sanskrit text reports that Buddha enumerated numbers up to 10^{53}. He also described six more numbering systems above these, leading to a number equivalent to 10^{421}. This was the highest number the ancients could think up and was as close to infinity as they could reach.

The Jain mathematicians recognized five different types of infinities as early as the third or second century BC—infinity in one direction, in two directions, in area, infinity everywhere and perpetual infinity.

Indians, like all other cultures, gave names to their numbers. The smaller numbers had numerous names each and these were derived from philosophical and spiritual concepts.

Zero was named 'shunya'. But there were other names describing zero—akasha, ambara, vyoman, all meaning the sky. The number 'one' was named 'eka' which denotes indivisibility. Two was named 'dvi' after Ashwini, the twice born. The number three was denoted by 'tri' after the three Vedas, Shiva's three eyes, his trident and so on. The number five was named 'pancha' after the five elements, the Pandavas or Rudra's five faces. The number nine was named 'nava' after the nine planets and the nine jewels. And the number ten was called 'dasha' after Vishnu's ten avatars or Ravana's ten heads.

I included this chapter to show you how rich our cultural heritage is. Most of the mathematical concepts that we know and use appear to have come from the west. But Indians had discovered many of these ideas long ago and had lost this knowledge due to various geographical and political reasons.

20

DO NUMBERS EXIST?

That sounds like a pretty dumb question, doesn't it? We all know what numbers are and what we can do with them and so on. However, we have philosophers who ponder over questions like these. The word 'philosophy' is of Greek origin. It means 'love of wisdom'. Philosophy consists of different branches of thinking that try to answer some of the basic questions about human life—for example, are human beings basically good or evil, and so on.

Philosophers are *deep* thinkers. They are not satisfied until they know a subject in its totality. Philosophers who study Mathematics would like to know the assumptions the subject is built on. They would like to have a clear understanding of the methods used in Mathematics. They would also want to know what Mathematics implies and understand its place in people's lives.

Since the subject is quite abstract, different people have arrived at different conclusions. Some of these conclusions can seem bizarre, some funny and some completely outlandish! Let's take a look at some of the conclusions they have reached.

Platonism or realism is a school of thought that states that numbers and other mathematical entities (for example, geometrical shapes and symbols related to Mathematics) actually exist. They are abstract and independent of all human activity. Numbers exist outside of time and space, independent of the human mind. They exist in a realm, which we access when we think of them! Human beings do not invent mathematical concepts. Rather, they discover them. Triangles, for example, are real entities and not creations of the human mind. Mathematicians happened to discover triangles and their properties.

The argument against Platonism is that the theory is too far-fetched. And what could be the realm that the platonists are speaking of?

Nominalism claims that mathematical objects do not exist as abstract entities just for us to make sense of them, and that Mathematics is only about objects in this world. For example, we say that there are 2 pencils, and so on. The 2 refers to pencils, which are concrete objects. Multiplication would be explained as follows:

6 lots of 7 objects would contain 42 objects. This is the reason that $6 \times 7 = 42$. However, nominalism runs into trouble when we discuss imaginary numbers.

Empiricists believe that Mathematics is like Science. We have to discover laws and we can't make statements without some proof. The argument against empiricism is that sometimes, in Science, one may arrive at the wrong conclusion by mistake. But in Mathematics, one rarely goes wrong.

Mathematical monism takes platonism one step further. It says that not only do numbers exist, numbers are the only things that exist. In fact, the basis of the universe is Mathematics! Imagine a computer with a gigantic memory and an astonishing processing speed. If we can put in all the information about position, momentum and all the properties of the fundamental particles present in the universe, this computer will be able to create the universe inside it. From the philosopher's point of view, this simulation would look unreal. But for the creatures that the computer has generated (including us human beings), this world would feel real. We are just mathematical structures that have become aware of ourselves! Sounds a bit like science fiction, doesn't it?

For further information on this, read Douglas Adams' five-part series *The Hitchhiker's Guide to the Galaxy* (and remember the number 42!) or watch *The Matrix* trilogy (they stopped at three; they were respectful of numbers like 3).

Now let us see what logicism has to say about numbers and Mathematics. Logicists believe that Mathematics can be derived entirely from logic. What is logic? Logic is correct reasoning. According to this theory, there are no mathematical objects that exist, or rules to be discovered. If one uses correct reasoning and a few basic assumptions, all of Mathematics can be formulated. That's not so exciting, is it?

Formalism, which is another school of thought, would have us believe that there is no deeper meaning to Mathematics and that the world would exist very well without it. Mathematics is just a game that uses certain symbols and rules. It just helps us quantify the world. But beyond that, it's just a mental exercise.

Conventionalism is a viewpoint that would have us believe that the rules of Mathematics are somewhat like the rules of grammar. One could have had alternative, equally workable rules that would have served the purpose.

Psychologism is another school of thought. Philosophers who believe in it hold the view that the principles of Maths have nothing to do with the world. They exist only in the human mind.

Intuitionists hold a similar viewpoint. They believe that Mathematics is the result of constructive mental activity of human beings and have little to do with reality.

There are many more schools of thought on the basis of Mathematics. Out of these, fictionalism is probably the most entertaining. According to fictionalism, all mathematical discourse is helpful, but false. Numbers don't exist. It's all just a useful story but has no basis in reality. For example, the Bible has stories that are useful and point to certain truths, but one need not believe in them! The same goes for Mathematics.

As we can see, philosophers have spent a lot of time and thought trying to answer questions that probably cannot be answered. That is because the world of numbers is mysterious. When we have dealt with numbers for some time, they do seem real to us. Each number seems to have a character of its own. We even think of certain numbers as lucky and some as unlucky. But can we touch numbers or feel them? We can't. They remain illusory, intangible and something to ponder on.

FURTHER READING

When you are older (or even now, if you really love numbers), you may want to read these books (in no particular order). They are all exciting and you don't have to be a math genius to understand them.

1. *The Colossal Book of Mathematics* – Martin Gardner
2. *Magical Mathematics: The Mathematical Ideas That Animate Great Magic Tricks* – Persi Diaconis and Ron Graham
3. *Zero: The Biography of a Dangerous Idea* – Charles Selfe
4. *The Joy of X: A Guided Tour of Mathematics, from One to Infinity* – Steven Strogatz
5. *Things to Make and Do in the Fourth Dimension: A Mathematician's Journey through Narcissistic Numbers, Optimal Dating Algorithms, At Least Two Kinds of Infinity, and More* – Matt Parker
6. *The Number Devil: A Mathematical Adventure* – Hans Magnus Enzensberger
7. *Professor Stewart's Cabinet of Mathematical Curiosities* – Ian Stewart

8. *A History of Pi* – Petr Backmann
9. *Journey through Genius: The Great Theorems of Mathematics* – William Dunham
10. *The Man Who Knew Infinity: A Biography of Srinivasa Ramanujam* – Robert Kanigel
11. *Godel Escher Bach: An Eternal Golden Braid* – Douglas Hofstadter

ACKNOWLEDGEMENTS

I wish to thank my father who taught me to love Mathematics and my mother who taught me to stay focused in life. My husband Sandipan Deb helped me select the topics in the book. My daughter Sukanya Deb painstakingly went through the entire manuscript and came up with useful suggestions. I am grateful to both of them.